Everyday Mathematics®

The University of Chicago School Mathematics Project

STUDENT MATH JOURNAL

VOLUME 2

McGraw Hill Education

The University of Chicago School Mathematics Project

Max Bell, Director, *Everyday Mathematics* First Edition; James McBride, Director, *Everyday Mathematics* Second Edition; Andy Isaacs, Director, *Everyday Mathematics* Third, CCSS, and Fourth Editions; Amy Dillard, Associate Director, *Everyday Mathematics* Third Edition; Rachel Malpass McCall, Associate Director, *Everyday Mathematics* CCSS and Fourth Editions; Mary Ellen Dairyko, Associate Director, *Everyday Mathematics* Fourth Edition

Authors
Max Bell, John Bretzlauf, Amy Dillard, Robert Hartfield, Andy Isaacs, James McBride, Kathleen Pitvorec, Peter Saecker, ‡Sarah R. Burns, *Ann McCarty, Robert Balfanz, †William Carroll

*Third Edition only
†First Edition only
‡Common Core State Standards Edition only

Fourth Edition Grade 6 Team Leader
Kathleen Pitvorec

Writers
Jorge Abner Bardeguez-Delgado, Kelly Darke, Cathy Hynes Feldman, Jennifer L. Jankowski, Soundarya Radhakrishnan

Differentiation Team
Ava Belisle-Chatterjee, Leader; Jean Capper, Martin Gartzman, Barbara Molina

Digital Development Team
Carla Agard-Strickland, Leader; John Benson, Gregory Berns-Leone, Juan Camilo Acevedo

Virtual Learning Community
Meg Schleppenbach Bates, Cheryl G. Moran, Margaret Sharkey

Technical Art
Diana Barrie, Senior Artist; Cherry Inthalangsy

UCSMP Editorial
Lila K.S. Goldstein, Senior Editor; Serena Hohmann, Rachel Jacobs, Kristen Pasmore, Delna Weil

Field Test Coordination
Denise A. Porter, Angela Schieffer, Amanda Zimolzak

Field Test Teachers
Jason Antesbeger, Kristin M. Arras, Catherine Ditto, Benjamin Kovacs, Joshua Ryan Marburger, Dawn A. Meziere, Kyle Radcliff, Lauren Scherer, Sara Sharp

Digital Field Test Teachers
Colleen Girard, Michelle Kutanovski, Gina Cipriani, Retonyar Ringold, Catherine Rollings, Julia Schacht, Christine Molina-Rebecca, Monica Diaz de Leon, Tiffany Barnes, Andrea Bonanno-Lersch, Debra Fields, Kellie Johnson, Elyse D'Andrea, Katie Fielden, Jamie Henry, Jill Parisi, Lauren Wolkhamer, Kenecia Moore, Julie Spaite, Sue White, Damaris Miles, Kelly Fitzgerald

Contributors
John Benson, Kelley E. Buchheister, Kathryn B. Chval, Andy Carter, James Flanders, Lila K.S. Goldstein, Aaron T. Hill, Serena Hohmann, Jeanne Mills DiDomenico, Denise Porter, Kathryn M. Rich, Mollie Rudnick, Sheila Sconiers, Laurie K. Thrasher, Penny Williams

Center for Elementary Mathematics and Science Education Administration
Martin Gartzman, Executive Director; Meri B. Fohran, Jose J. Fragoso, Jr., Regina Littleton, Laurie K. Thrasher

External Reviewers
The *Everyday Mathematics* authors gratefully acknowledge the work of the many scholars and teachers who reviewed plans for this edition. All decisions regarding the content and pedagogy of *Everyday Mathematics* were made by the authors and do not necessarily reflect the views of those listed below.

Elizabeth Babcock, California Academy of Sciences; Arthur J. Baroody, University of Illinois at Urbana-Champaign and University of Denver; Dawn Berk, University of Delaware; Diane J. Briars, Pittsburgh, Pennsylvania; Kathryn B. Chval, University of Missouri–Columbia; Kathleen Cramer, University of Minnesota; Ethan Danahy, Tufts University; Tom de Boor, Grunwald Associates; Louis V. DiBello, University of Illinois at Chicago; Corey Drake, Michigan State University; David Foster, Silicon Valley Mathematics Initiative; Funda Gönülateş, Michigan State University; M. Kathleen Heid, Pennsylvania State University; Natalie Jakucyn, Glenbrook South High School, Glenview, IL; Richard G. Kron, University of Chicago; Richard Lehrer, Vanderbilt University; Susan C. Levine, University of Chicago; Lorraine M. Males, University of Nebraska-Lincoln; Dr. George Mehler, Temple University and Central Bucks School District, Pennsylvania; Kenny Huy Nguyen, North Carolina State University; Mark Oreglia, University of Chicago; Sandra Overcash, Virginia Beach City Public Schools, Virginia; Raedy M. Ping, University of Chicago; Kevin L. Polk, Aveniros LLC; Sarah R. Powell, University of Texas at Austin; Janine T. Remillard, University of Pennsylvania; John P. Smith III, Michigan State University; Mary Kay Stein, University of Pittsburgh; Dale Truding, Arlington Heights District 25, Arlington Heights, Illinois; Judith S. Zawojewski, Illinois Institute of Technology

Note
Many people have contributed to the creation of *Everyday Mathematics*. Visit http://everydaymath.uchicago.edu/authors/ for biographical sketches of *Everyday Mathematics* Fourth Edition staff and copyright pages from earlier editions.

www.everydaymath.com

Send all inquiries to:
McGraw-Hill Education
8787 Orion Place
Columbus, OH 43240

ISBN: 978-0-02-135252-4
MHID: 0-02-135252-6

Printed in the United States of America.

3 4 5 6 7 8 9 QSX 20 19 18 17 16

Contents

Unit 6

Unit 7

Unit 8

Activity Sheets

Math Boxes

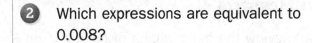

---

① Evaluate.

a. $9^2 =$ _____

b. $5^3 =$ _____

c. $0^5 =$ _____

d. $\left(\frac{1}{4}\right)^2 =$ _____

e. $0.2^2 =$ _____

SRB 98

---

② Which expressions are equivalent to 0.008?

Circle ALL that apply.

A.   8%

B. $\frac{8}{1,000}$

C.   0.8%

D. $\frac{8}{100}$

SRB 56-58

---

③ Here are the ages, in years, of the dancers in a beginning ballet class:

8, 9, 9, 10, 11, 11, 12, 13

Find the mean and median ages.

Mean: _____

Median: _____

SRB 284

---

④ Compare with >, <, or =.

a. 1.8 _____ −1.8

b. |−1.8| _____ −1.8

c. $\frac{1}{8}$ _____ 1.8

d. $-\frac{1}{8}$ _____ |−1.8|

SRB 92-93

---

⑤ Aileen has $5\frac{1}{3}$ yards of fabric. She needs $1\frac{1}{3}$ yards of fabric to make the skirt she designed. How many skirts can Aileen make with the fabric she has?

Number model: _____

Solution: _____

SRB 32

---

⑥ Use the Distributive Property to write an equivalent expression.

a. $3 * (8 - 5) =$ _____

b. _____ $= 18 + 48$

c. $1.5 * 99 =$ _____

SRB 204-205

217

Polygons on the Coordinate Grid

Math Message

 a. Review the descriptions of polygons on *Student Reference Book*, page 238. Find polygons in the faces of objects around the classroom. Sketch or list at least four objects.

b. List the names of some of the polygons you found.

_____

_____

On the coordinate grid, draw the picture of Elektro using the clues for points *A* through *Y* below. Do the following:

- For each point, find any missing coordinates.

- Plot and label the point on the grid.

- Use a straightedge to connect points when instructed to do so.

2   *A*: (−8, 8)   *B*: (−5, 11)   *C*: (−2, 11)   *D*: (_____, _____)

Point *D* has the same *y*-coordinate as point *A*, but it is in a different quadrant. $\overline{AD}$ has a length of 9 units.

Connect the points to form polygon *ABCD*. What is the shape of *ABCD*? _____

3   *E*: (_____, _____) Point *E* has the same *y*-coordinate as point *A*.
The length of $\overline{ED}$ can be represented with the number model $|-2| + |1|$.

F: (_____, _____) Point *F* has the same *y*-coordinate as point *E*, and it is between points *A* and *E*. $\overline{AF}$ has a length of 3 units.

G: (_____, _____) Polygon *FEGH* is a square.

H: (_____, _____)

Connect the points to draw square *FEGH*.

218

4 I: (_____ , _____) Point I is closer to point H than to point G.
The length of $\overline{IH}$ can be represented with the number model $|-9| - |-5|$.

J: (_____ , _____) Point J is in a different quadrant than point G.
The length of $\overline{GJ}$ can be represented by the number model $|-2| + |2|$.

K: (_____ , _____) The x-coordinates of points J and K are the same.
The length of $\overline{JK}$ can be represented by the number model $|-7| + |5|$.

L: (_____ , _____) Point L is the fourth vertex of rectangle $IJKL$.

Connect the points to draw rectangle $IJKL$.

5 M: (_____ , 2.5) Point M is on $\overline{JK}$.

N: (6.5, 5)

O: (_____ , _____) The length of $\overline{KO}$ can be represented by $|-7| - |-2|$.

Connect the points to draw triangle MNO.

6 P: (−9, −2) R: (−13, −2) Q: (−9, _____)

The length of $\overline{PQ}$ is 4.5 units. Point P is in a different quadrant than point Q.

Connect the points to draw triangle PQR.

7 S: (−7.5, −7) T: (−4, _____)

Point T is on $\overline{LK}$.

U: (−4, _____) $\overline{TU}$ is 5 units long. Point T is closer to point U than to the x-axis.

V: (_____ , _____) V is the fourth vertex of rectangle $STUV$.

Connect the points to draw rectangle $STUV$.

8 W: (_____ , _____) Point W is on $\overline{LK}$ and the length of $\overline{LW}$ is 7.5 units.

X: (7.5, −12)

Y: (_____ , _____) Point Y is the fourth vertex of parallelogram $WKXY$.

Connect the points to draw parallelogram $WKXY$.

Polygons on the Coordinate Grid (continued)

Follow the directions to find the lengths of Elektro's line segments.

9 Consider $\overline{QP}$.

    a. What are the coordinates of the endpoints of $\overline{QP}$?

    (_____ , _____) and (_____ , _____)

    b. Write an expression for calculating the length of $\overline{QP}$.   _____

    c. How long is $\overline{QP}$? _____

10 Consider $\overline{IG}$.

    a. What are the coordinates of the endpoints of $\overline{IG}$?

    (_____ , _____) and (_____ , _____)

    b. Write an expression for calculating the length of $\overline{IG}$.   _____

    c. How long is $\overline{IG}$? _____

11 **a.** If you connected Points V and X, what would be the coordinate of the point where $\overline{VX}$ intersects the y-axis?   _____

    b. If you extended $\overline{BC}$, what would be the coordinates of the point where the line segment intersects the y-axis?   _____

    c. Write an expression for calculating Elektro's height.   _____

    d. How tall is Elektro? _____

Try This

12 **a.** What would be the coordinates for Elektro's head after applying the rule (0.5x, 0.5y)?

    _____

    b. Write a number model and find the length of the new $\overline{AD}$. _____

    c. Compare the length of the new $\overline{AD}$ to the length of the original $\overline{AD}$ given in Problem 2.

    Explain how you know the new length makes sense.

    _____

    _____

Finding Areas of Polygons with Rectangles

Math Message

1 For each polygon below, show and explain how you can use the areas of rectangles to help you find the area of the polygon.

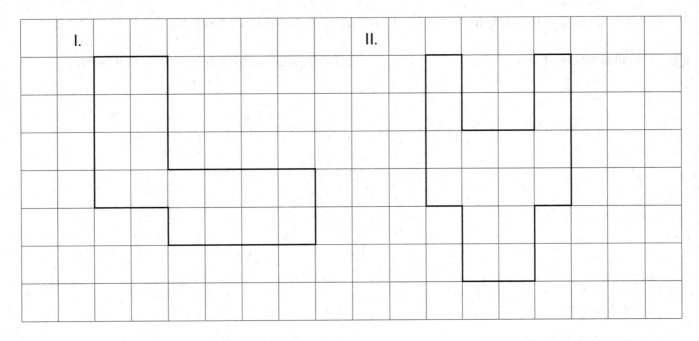

I. _____

_____

_____

_____

_____

II. _____

_____

_____

_____

_____

2 Examine the examples below and sketch the height for Parallelogram V.
Label the base and height for Parallelogram V.

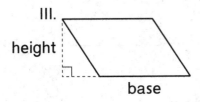

III.

height

base

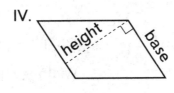

IV.

height

base

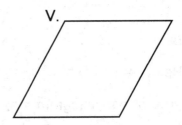

V.

Areas of Parallelograms

For Problems 1–3, cut out Parallelograms A–C on *Math Masters*, page 208.

DO NOT CUT OUT THE ONES BELOW.

Cut each parallelogram into two pieces so that it can be made into a rectangle.

1 cm²

Draw line segments on Parallelograms A and B below to show their heights.

① **Parallelogram A**

Tape your rectangle in the space below.

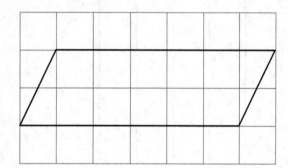

Base = _____   Length = _____

Height = _____   Width (height) = _____

Area of parallelogram = _____   Area of rectangle = _____

② **Parallelogram B**

Tape your rectangle in the space below.

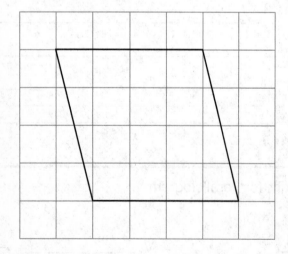

Base = _____   Length = _____

Height = _____   Width (height) = _____

Area of parallelogram = _____   Area of rectangle = _____

3 Draw a line segment outside Parallelogram C to show its height.

Parallelogram C   Tape your rectangle in the space below.

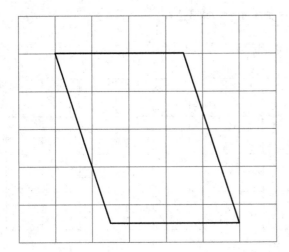

Base = _____   Length = _____

Height = _____   Width (height) = _____

Area of parallelogram = _____   Area of rectangle = _____

4 **a.** Look for patterns in Problems 1–3. Use the patterns you find to write a formula for the area of a parallelogram.

b. Use your formula to find the area of parallelogram *DORA*.

Use your ruler to measure where needed.

Draw on and label the parallelogram to show what you measured.

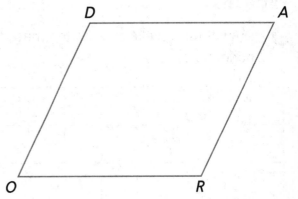

Area of parallelogram *DORA* _____

c. Explain how you used the formula to find the area for parallelogram *DORA*.

_____

_____

5   For Parts a, b, and c, draw the polygon on the grid and label the height (*h*) and base (*b*):

a. A rectangle whose area is 12 cm²

b. A parallelogram that is not a rectangle and has an area of 12 cm²

c. A different nonrectangular parallelogram with an area of 12 cm²

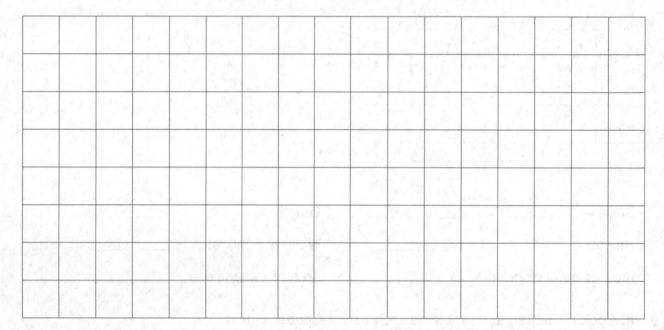

Try This

6   Draw three parallelograms that have the same base and the same area but different perimeters.

Greatest Common Factor and Least Common Multiple

1 Find the greatest common factor.

    a. GCF (64, 88) = _____

    b. GCF (30, 75) = _____

    c. GCF (27, 35) = _____

    d. GCF (18, 24, 48) = _____

2 Find the least common multiple.

    a. LCM (3, 11) = _____

    b. LCM (12, 5) = _____

    c. LCM (8, 12) = _____

    d. LCM (10, 5, 6) = _____

3 Find a pair of numbers that works for each set of clues.
Hint: Use the GCF/LCM grid method.

    a. The LCM is 48. The GCF is 12.

    b. The LCM is 54. The GCF is 9.
       One of the numbers in the pair is NOT 9.

4 Sharon can buy cards in boxes of 12 and stamps in packages of 20.
She wants the number of cards and stamps to match exactly.
What is the least number of boxes of cards she should buy? _____

How many packages of stamps will she need for the cards? _____

5 A box has 48 red marbles and 72 blue marbles.
They are divided into identical piles.
(For example, there could be two piles of 24 red marbles and 36 blue marbles each.)

What is the greatest number of piles that can be made? _____

How many of each color are in one pile? _____

Math Boxes

1 Alice's snack mix has $10\frac{1}{2}$ cups of raisins.

How many cups of
pretzels should it have? _____

Use the ratio/rate table to help you
answer the question.

Cups of Raisins	3				
Cups of Pretzels	5				

SRB
43-44

2 Evaluate.

a. _____ $= 53 - 4 * 9 + 8 * 2$

b. $3 + 5^2 * 3 =$ _____

c. $(5.2 + 3) * (6.4 - 1.4) =$ _____

SRB
203

3 Estimate and then calculate.

a. $1.2 * 10.5$

Estimation number sentence:

_____

Answer: _____

b. $2.34 * 6.9$

Estimation number sentence:

_____

Answer: _____

SRB
133-138

4 Find the mean absolute deviation (m.a.d.)
for the five test scores below.

80, 84, 72, 93, 81

m.a.d. _____

SRB
294

5 **Writing/Reasoning** How can you use the mean absolute deviation as a tool for
understanding the variability of data?

226

Exploring Triangles

Math Message

1 Work with a partner.

    a. Cut apart one set of triangle cards from *Math Masters*, page 212.

    b. Circle how you will sort the triangles: By side length   By angle measure

    c. Sort the triangle cards into 3 piles according to your chosen attribute (sides or angles).

    d. For each group, list the triangles that belong in the group and briefly describe the
        rule for the group.

        Group 1 Rule: _____

                Triangles: _____

        Group 2 Rule: _____

                Triangles: _____

        Group 3 Rule: _____

                Triangles: _____

---

2 Cut out Triangles A and B from *Math Masters*, page 213.
DO NOT CUT OUT THE TRIANGLE BELOW.
Tape Triangles A and B together to form a parallelogram.

1 cm²

Triangle A

Tape your parallelogram in the space below.

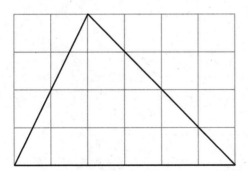

Base = _____

Height = _____

Area of triangle = _____

Base = _____

Height = _____

Area of parallelogram = _____

Exploring Triangles (continued)

3 Cut out Triangles C and D from *Math Masters,* page 213.
Draw a line segment to indicate the height of Triangle C below.
Tape Triangles C and D together to form a parallelogram.

Triangle C

Tape your parallelogram in the space below.

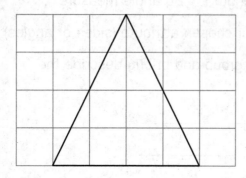

Base = _____

Base = _____

Height = _____

Height = _____

Area of triangle = _____

Area of parallelogram = _____

4 Cut out Triangles E and F from *Math Masters,* page 213.
Draw a line segment to indicate the height of Triangle E below.
Tape Triangles E and F together to form a parallelogram.

Triangle E

Tape your parallelogram in the space below.

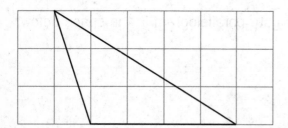

Base = _____

Base = _____

Height = _____

Height = _____

Area of triangle = _____

Area of parallelogram = _____

Exploring Triangles (continued)

5. Look for patterns in Problems 2–4. Use the patterns you find to write a formula for the area of a triangle. Be sure to define your variables.

_____

Use triangle *TRI* for Problems 6–7.

6. Measure and label one base and one height for triangle *TRI*. Measure to the nearest tenth of a centimeter. Use your formula to calculate the approximate area for triangle *TRI*.

Number sentence: _____

Approximate area of triangle *TRI*: _____

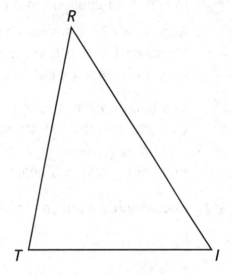

7. **a.** Label the other bases and heights for triangle *TRI*.
   Measure each base and height to the nearest tenth of a centimeter.
   Complete the table with approximate measurements and calculated areas.

Side Used as Base	Base (cm)	Height (cm)	Area of Triangle (cm²)
TI			
TR			
RI			

b. The areas you found in Part a should be about the same.
   What might cause the variation in your calculated areas?

_____

_____

229

Reviewing Inequalities

1 For each statement, circle the number of solutions that will make it true.

    a. $105 + c = 478$ None   One   Many

    b. $105 + c > 478$ None   One   Many

2 Match each situation to a number model.

Rudy had 478 stamps in his collection after 5 years.
He started with 105 stamps that his brother Bob gave him.
How many stamps did he collect during the 5 years?

                                                     $105 + c = 478$

Bob had duplicates of the 105 stamps he gave Rudy.
Bob started with 105 stamps and collected more than
Rudy during the same 5 years. How many stamps did
Bob collect during 5 years?

                                                     $105 + c > 478$

3 Record three solutions for each inequality and three numbers that are not solutions.

Inequality	Solutions	Non-solutions
a. $235 > x$	_____	_____
b. $y + 5 \leq 10$	_____	_____
c. $12 \geq 2d$	_____	_____

4 Graph the solutions to each inequality.

    a. $p \leq 4.5$

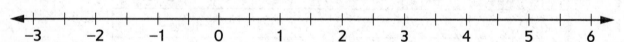

    b. $-3\frac{1}{2} > q$

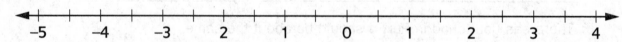

5 Write an inequality for each situation.

    a. n is less than double 7.5   _____

    b. $25\frac{1}{2}$ is greater than p added to 100   _____

    c. $\frac{1}{2}$ of 5,048 is greater than or equal to m _____

Math Boxes

---

① Write each expression using an exponent.

a. $12 * 12 =$ _____

b. $2 * 2 * 2 * 2 =$ _____

c. $1 * 1 * 1 * 1 * 1 =$ _____

d. $1.3 * 1.3 =$ _____

e. $a * a * a =$ _____

SRB
98

---

② Circle answers that are equivalent to $\frac{3}{100}$.

A.   3%

B.   30%

C.   300%

D.   0.03

SRB
56-58

---

③ Find the landmarks for the data set.

34, 35, 33, 37, 37, 40, 30, 36,
40, 39, 47, 36

Mean: _____

Median: _____

Describe a situation for the data set.

_____

SRB
284

---

④ Compare with >, <, or =.

a. 12.48 _____ -12.48

b. $|-12.48|$ _____ -12.48

c. $\frac{25}{10}$ _____ 1.48

d. $-\frac{25}{10}$ _____ $|-1.48|$

SRB
92-93

---

⑤ Lamar has $\frac{5}{6}$ of an hour of free time. It takes him $\frac{1}{3}$ of an hour to play his favorite video game. How many whole games does he have time to play?

Number model: _____

Solution: _____

SRB
32

---

⑥ Use the Distributive Property to write an equivalent expression.

a. $12 * 68 =$ _____

b. _____ $= 88 + 55$

c. $42.5 * 99 =$ _____

SRB
204-205

Composing and Decomposing Polygons

Math Message

1 Amara enjoys tangram puzzles. Her mom said she could paint two puzzles on a bedroom
wall. The pictures below are scale drawings of the puzzles she will paint on her wall.
Draw line segments on the scale drawings to show the tangram pieces that make each
puzzle. Calculate the total area each puzzle will cover on Amara's wall. Use the
measurements listed and tangram relationships to find other measurements you need.

Shape 1   **Shape 2**

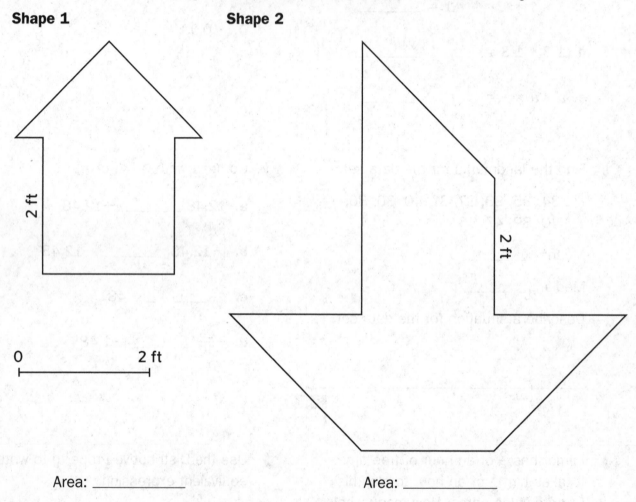

Area: _____   Area: _____

2 Amara's mom said the paintings could not cover more than 25% of the wall.
The rectangular wall is 12 feet long and 8 feet high.

a. About what percent of the wall area will the painted puzzles cover? _____

b. Explain how you solved the problem. _____

_____

Composing and Decomposing Polygons (continued)

3   Estimate the area of the state of Nevada.

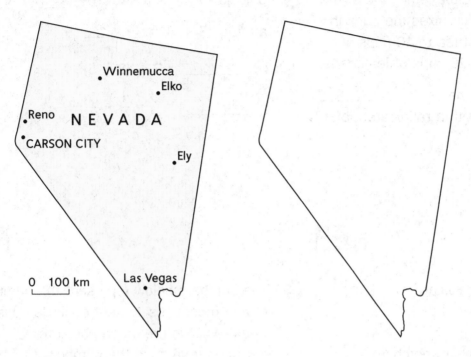

a. Use a ruler. Draw line segments to show how to **decompose** the state into polygons to calculate the approximate area.

Measure line segments that you will use to the nearest tenth of a centimeter.

Use the map scale to figure out distances.

Calculate and label the areas of the polygons.

b. Approximate area of Nevada: _____

c. Explain how you decomposed Nevada into polygons, and describe how you found its total area.

_____

_____

_____

_____

_____

Math Boxes

1 Andre is making iced-tea lemonade at the school fund-raiser. He mixes the drink in a ratio of 18 cups of tea to 10 cups of lemonade. He uses 25 cups of lemonade.

How much tea does he use? _____
Solve the problem with a ratio/rate table.

SRB
43-44

2 Calculate.

a. $15 - 7 + 3 =$ _____

b. _____ $= 9 \div (8 - 5)^2$

c. _____ $= \frac{2}{3} * \frac{3}{4} + \frac{5}{9} \div \frac{1}{3}$

SRB
203

3 Estimate and then calculate.

a. $103 * 5.2$

Estimation number sentence:

_____

Answer: _____

b. $0.07 * 10.9$

Estimation number sentence:

_____

Answer: _____

SRB
133-138

4 The numbers below represent the number of text messages Jo sent each day this week. What is the mean absolute deviation (m.a.d.) in the number of text messages Jo sent this week? Round your answer to the nearest hundredth.

8, 12, 3, 10, 6, 20, 18

m.a.d. _____

SRB
294

5 **Writing/Reasoning** Explain how the way you made your estimate helps you determine that your answer for Problem 3b makes sense.

Math Boxes

Creating and Comparing Nets

1   Circle the picture of the net you have been assigned to use.

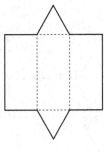 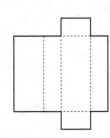 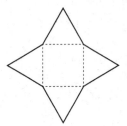 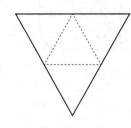

Triangular Prism   Rectangular Prism   Square Pyramid   Triangular Pyramid

2   Use the attributes of the geometric solid you built to answer the following questions:

a. Which polygons make up the faces of your solid? How many of each polygon are there?

_____

b. Which faces are parallel, if any? _____

c. Which faces are congruent, if any? _____

d. How many edges are there? How many vertices? _____

3   Compare your geometric solid with the solids other students made.
How are they alike and how are they different? Complete the table below.

Name	Number and Shape of Faces	Which Faces Are Congruent?	Number of Edges	Number of Vertices
Rectangular prism				
Triangular prism				
Square pyramid				
Triangular pyramid				

Matching Solids and Nets

1 Draw lines to connect the pictures of geometric solids with the nets that represent them.

A.

M.

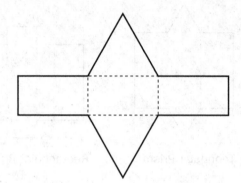

B.

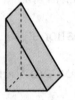

N.

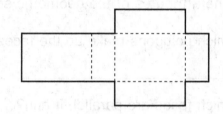

O.

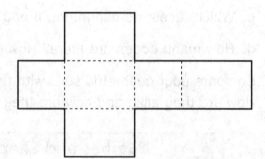

C.

D.

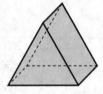

P.

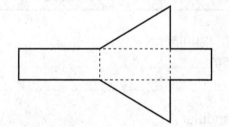

2 Record one of the matches you made in Problem 1, and explain how you know you could fold the net to make the solid.

_____

_____

_____

_____

Math Boxes

① Graph the inequalities.

a. $-4 > t$

```
 ←——+——+——+——+——+——+——+——→
   -6  -5  -4  -3  -2  -1   0   1
```

b. $k \leq 2.5$

```
 ←——+——+——+——+——+——+——+——→
   -3  -2  -1   0   1   2   3   4
```

SRB
210-211

② Record two ordered pairs with the same *x*-coordinate that are in different quadrants.

_____

Explain how you know they are in different quadrants.

_____

_____

_____

SRB
95,
265-266

③ Write an algebraic expression to represent each statement.

a. 5 more than *f* _____

b. 10 less than *g* _____

c. *g* less than 10 _____

d. half of *p* _____

SRB
200-202

④ Fill in the missing digits in this multiplication algorithm.

```
              5   3   9
    *             6   7
    ———————————————————
          3   7  □   3
  + □   □      3  □   □
    ———————————————————
      3  □ , □   1   3
```

SRB
137-138

⑤ **Writing/Reasoning** Explain how you can use partial products to check that your answers for Problem 4 make sense.

237

Exploring Surface Areas

Math Message

1 On the grid, draw a net for the pictured prism.

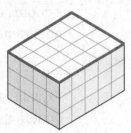

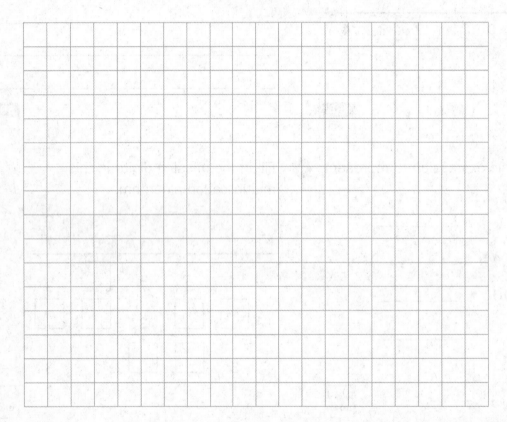

2 Find the outlined face on the prism in Problem 1. Shade the same face in your drawing.

3 The prism in Problem 1 is a scale drawing of a kindergarten building block. A kindergarten teacher plans to paint the block. Each square on the scale drawing represents a 1" × 1" square.

How many square inches will the paint cover? _____

4 Explain how the drawing of the net could help you calculate the amount of paint needed.

_____

_____

Solving a Packaging Problem

Anita designs packaging for a company that sells roasted cashews.
She designed two different boxes the company might use to package cashews.
Below are her two designs. They have about the same volume.

Design 1

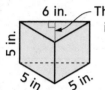

6 in. — The height of the triangle
in the base is 4 inches.

5 in.

5 in.   5 in.

Design 2

The height of each —
triangle is 5 inches.

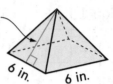

6 in.   6 in.

1. Use the listed measurements to draw a net for each design on the $\frac{1}{4}$-inch grid on *Math Masters*, page TA17. The length of one square on the grid represents 1 inch.

2. Sketch a net for each design below.

   Design 1   **Design 2**

3. Use your nets and number sentences to find the **surface area** of each design. The surface area tells how much cardboard each design requires.

   Design 1   **Design 2**
   Number sentence: Number sentence:

   _____   _____

   Cardboard required: _____   Cardboard required: _____

4. Besides the amount of material each design requires, what else might the company consider when picking a design?

   _____

   _____

239

Surface Area of a Package

1   Anita decided to explore the amount of cardboard needed to make cube-shaped packages
    of different sizes.

	Edge Length	Number Sentence	Surface Area
Cube Design 1	10 in.		
Cube Design 2	20 in.		
Cube Design 3	5 in.		
Cube Design 4	40 in.		

2   Describe the relationship between the edge length of a cube and the total surface area you
    found for the cubes in Problem 1.

    _____

    _____

    _____

3   Write an algebraic expression you could use
    to find the surface area of a cube with edge length *n*. _____

Try This

4   The surface area of a cube is 96 cm².
    Sketch a net for the cube and label one of the edge lengths.

Practicing with Tape Diagrams

Here is a tape diagram showing the number of baseball cards two friends have.
Use it to solve Problems 1–3.

Carol	28	28			
Fin	28	28	28	28	28

1 What is the ratio of Carol's cards to Fin's? _____

2 How many cards does Carol have?   _____

3 The two friends got a large box of additional cards to share.
The ratio stays the same, but the total number of cards is now 308.

How many cards does Carol have now?   _____

For every 3 text messages Sam sends, Ben sends 4. Use tape diagrams to solve Problems 4–6.

4 Last week Sam sent 36 messages.
How many did Ben send?

_____

5 Last month Sam and Ben
sent 350 messages in total.
How many did Sam send?

_____

6 Sam sends a total of
225 messages one month.
How many does Ben send?

_____

Math Boxes

1 South Australia's Lake Eyre is 52 feet below sea level. Tanzania's Mount Kilimanjaro is 19,341 feet above sea level. How far apart are the levels of these two locations?

Number model: _____

Solution: _____

SRB
92-83

2 **a.** Use mathematical language to describe the expression $5 * 4^2$.

_____

_____

b. Evaluate: $5 * 4^2 =$ _____

SRB
200-202

3 Complete the five-number summary for this box plot.

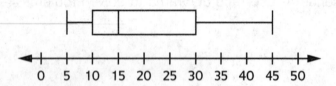

0 5 10 15 20 25 30 35 40 45 50

Minimum: _____ Q1: _____ Median: _____ Q3: _____ Maximum: _____

SRB
300-302

4 Multiply or divide.

a. $\frac{3}{4} \div \frac{3}{8} =$ _____

b. $\frac{8}{9} \div \frac{2}{3} =$ _____

c. _____ $= 1\frac{3}{4} * 3$

SRB
189,
193-195

5 An 18.5-pound bag of dog food sells for $21.99, and a 32-pound bag sells for $34.99. Use unit rates to determine which bag has the lower cost per pound. Which is the better buy? Explain your reasoning.

_____

_____

_____

_____

_____

SRB
32

Solving Surface-Area Problems

Math Message

The Rubik's Cube is a puzzle that starts with all of the colors mixed up. The goal is to realign the cube so that each face is all one color.

The world's largest Rubik's Cube was made for the Knoxville, Tennessee, World's Fair of 1982.

Each edge of the largest Rubik's Cube is 3 m long.

1 For the world's largest Rubik's Cube, approximately how many square meters will be painted

if the entire cube is painted? _____

---

2 Antoine's favorite mechanical pencil leads come in a container shaped like a triangular prism.

The base of each triangular face is about 2 cm long.

The height of each triangular face is about 1.7 cm long.

The container is about 6 cm long.

a. Label the container diagram with the measurements.

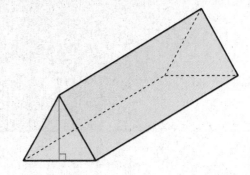

b. On the grid below, draw a net to represent the container. Scale: 1 grid square = 1 cm²

1 cm²											

c. Approximately how much plastic would it take to make the container? _____

Solving Surface-Area Problems (continued)

3 The Pyramide du Louvre in Paris is made of glass.

The structure is above the entrance to the Louvre Museum,
so the Pyramide has no base. It looks like a square pyramid.

The base of each triangular face is approximately 36 m.

The height of each triangular face is about 28 m.

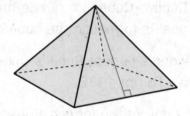

a. Label the pyramid diagram with the measurements.

b. On the grid below, draw a net to represent the Pyramide du Louvre.
(Include the base on the net even though it is not made of glass.)

Scale: 1 grid square = 4 m × 4 m (The length of each side is 4 m
and the area is 16 m².)

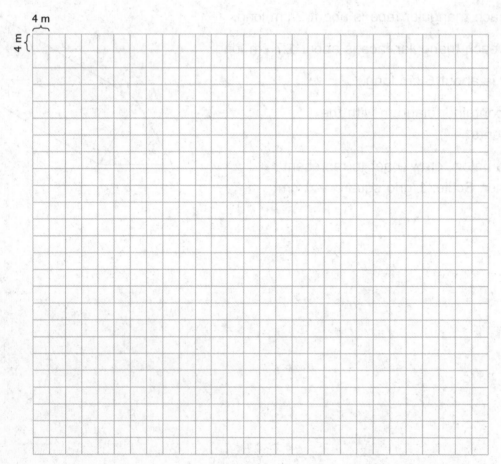

c. What is the area of each triangular face of the Pyramide du Louvre? _____

d. Approximately how many square meters of glass were used for the faces? _____

244

Math Boxes

1 Record three values that make each inequality true.

a. $n > -2$ _____

b. $k \leq 3\frac{1}{3}$ _____

c. $0 \geq m$ _____

SRB
210-211

2 Record two ordered pairs that have the same y-coordinate but are in different quadrants.

_____

Find the distance between them.

_____

_____

SRB
95,
265-266

3 Write an algebraic expression to represent each statement.

a. 8 more than half of f _____

b. 3 less than double g _____

c. triple m less than 100 _____

d. one quarter of p _____

SRB
200-202

4 Fill in the missing digits in this multiplication algorithm.

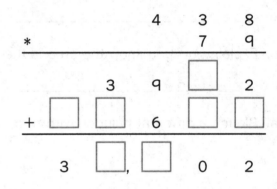

SRB
137-138

5 **Writing/Reasoning** For Problem 2, explain how you could use absolute value as a tool for finding the distance between your points.

Composing Polygons from Triangles

Math Message

Cut out several copies of Triangle 1 from *Math Masters*, page 224.
Use the copies of Triangle 1 as tools to help you solve the problems.
It is okay to cut apart and rearrange Triangle 1 to find your solutions.

1   **a.** Sketch a different polygon with the same area as Triangle 1.

     b. Explain how you used Triangle 1 to sketch a polygon with the same area.

     _____

     c. Explain how you know the area is the same.

     _____

2   **a.** Sketch a different polygon with twice the area of Triangle 1.

     b. Explain how you used Triangle 1 to sketch a polygon with twice the area.

     _____

Math Boxes

1 Argentina's Valdes Peninsula is 131 feet below sea level. Alaska's Mount McKinley is 20,320 feet above sea level. How far apart are the levels of these two locations?

Number model:

_____

Solution: _____

SRB
92-93

2 **a.** Use mathematical language to describe the expression $(19 - 4) * 3$.

_____

_____

_____

b. Evaluate: _____ $= (19 - 4) * 3$

SRB
200-202

3 Given the five-number summary below, draw the box plot.

Minimum: 60   Q1: 75   Median: 85   Q3: 90   Maximum: 100

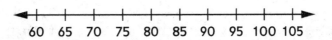

60 65 70 75 80 85 90 95 100 105

SRB
300-302

4 Estimate. Then multiply.

	Estimate	Answer
$1\frac{3}{4} * 3$		
$\frac{8}{9} * 1\frac{1}{2}$		
$3\frac{1}{2} * 1\frac{1}{7}$		

SRB
189,
193-195

5 One car used 15.5 gallons of gasoline to travel 372 miles. A second car traveled 198 miles using 7.2 gallons. Which car got better gas mileage?

Use rates to justify your answer.

Hint: Think about miles per gallon.

_____

_____

_____

SRB
32

Investigating the Volume of a Prism

Math Message

1 **a.** Examine the nets pictured on *Math Masters*, page 231.
Which prism do you think holds the most centimeter cubes? _____

b. Each group member should build one of the nets pictured.
Record the letter of the net you are building. _____

c. Predict about how many cubes you think your net will hold. _____

d. Explain how you made your prediction.

_____

_____

2 Cut out your net. Assemble your prism with tape. Leave one side of your prism open.
Gently pack your prism with centimeter cubes.
Line up the cubes so that you can fit as many as possible in your prism.

How many centimeter cubes fit in your prism? _____

3 Record the dimensions and volume of each prism from the Math Message.

Net	Length	Width	Height	Volume
A				
B				
C				
D				

4 Using *l* for the length of the base, *w* for the width of the base, and *h* for
the height, write a formula for the volume of a rectangular prism. $V =$ _____

5 Sometimes this formula is written as $V = Bh$.

a. What does *B* replace in the original formula? _____

b. What is the difference between what *b* in an area formula and *B* in a volume formula

represent? _____

_____

_____

Finding Volume

Here are the formulas for finding the volume of a prism: $V = l * w * h$ and $V = Bh$.
Use the formulas to find the volume of each prism two different ways.

1 First number model: _____

Volume: _____

Second number model: _____

Volume: _____

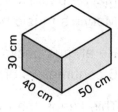

2 First number model: _____

Volume: _____

Second number model: _____

Volume: _____

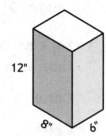

3 First number model: _____

Volume: _____

Second number model: _____

Volume: _____

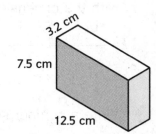

249

Shipping a Music Box

Marvin's Music Box ships fine music boxes to locations all over the world.
Music boxes are fragile and need to be packaged carefully.
Marvin packs each music box in packing peanuts.
Marvin's music boxes are basically shaped like rectangular prisms.

He always packs a music box so that he has room for *at least*
5 centimeters of packing peanuts all the way around the music box.

This keeps the music box safe in the shipping carton.

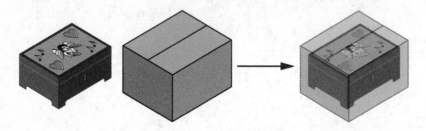

Marvin has an order for a music box that has the following dimensions:

Length: 20 cm   Width: 15 cm   Height: 10 cm

1 For each order, Marvin has two sizes of shipping cartons he can use.
Here are his shipping-carton choices:

Carton 1	Length: 50 cm	Width: 40 cm	Height: 30 cm
Carton 2	Length: 30 cm	Width: 25 cm	Height: 20 cm

Which carton should he use and why?

_____

_____

2 What volume of packing peanuts will he need to pack the music box in the carton? Assume
he will fill all of the empty space with peanuts. *Hint:* It may help to label the diagrams above
with the dimensions of the music box and the shipping carton you chose.

_____

_____

3 Write a number sentence to show how you solved Problem 2.

_____

Writing and Evaluating Algebraic Expressions

Remember that drawing a picture or diagram may help you solve a problem.
For each problem situation, do the following:

- Write an algebraic expression using the variable that is defined.

- Evaluate the expression for the given value of the variable.

1 The perimeter of a triangle with three sides of length x: _____

The perimeter when $x = 3.5$ inches: _____

2 The area of a square with side length v: _____

The area of a square when $v = 6\frac{1}{2}$ m: _____

3 Jorge is m years old. Jessica is 7 years younger than Jorge. _____

When $m = 22$ years, how old is Jessica? _____

4 The cost of s T-shirts that cost $12.99 each: _____

The cost of 6 T-shirts: _____

5 The cost per person when z people share a $154
bill for setting up a community garden: _____

The cost per person when $z = 8$: _____

6 The area of a rectangle with a length that is
4 units longer than its width of w: _____

The area when $w = 3.4$ cm: _____

Try This

7 The sum of three consecutive numbers,
when the first number is n: _____

What is the sum of all three consecutive numbers when $n = 11$? _____

Math Boxes
Preview for Unit 6

① Record whether each expression is true or false.

a. $|4| > |-3|$ _____

b. $200.03 > 200.030$ _____

c. $49 < |-49|$ _____

d. $72.056 < 72.5$ _____

SRB
92-93,
210-211

② Write the next three numbers in each pattern.

a. 5, 9, 13, 17, _____, _____, _____

b. 10, 30, 50, 70, _____, _____, _____

c. 100, 93, 86, 79, _____, _____, _____

SRB
221

③ Alexi has twice as many soccer practices as George. George has m soccer practices per month.

Algebraic expression: _____

How many soccer practices does Alexi have if George has 7 practices?

Number model: _____

Solution: _____

SRB
200-202

④ Evaluate.

a. $5x$, if $x = \frac{2}{5}$ _____

b. $5 - x$, if $x = 5$ _____

c. $2y - 1$, if $y = \frac{2}{3}$ _____

d. $h + \frac{3}{4}$, if $h = \frac{1}{2}$ _____

SRB
208-209

⑤ Find a value of x that makes this number sentence true: $\frac{3}{7}x = 1$.

Circle ALL that apply.

A. $x = 1$

B. $x = \frac{7}{3}$

C. $x = 7$

D. $x = 3$

SRB
208-209

⑥ For each expression, use the Distributive Property to write an equivalent expression.

a. $7 * (5 + \frac{1}{7}) =$ _____

b. $5 * (4 + z) =$ _____

c. $\frac{1}{7} * (28 + 84) =$ _____

SRB
204-205

Comparing Volume Units

Math Message

 a. On $\frac{1}{4}$-inch grid paper, draw a net to represent 1 cubic inch.
Hint: The number of grid squares each edge length spans is the number of $\frac{1}{4}$-inch increments in 1 inch.

b. What is the surface area of your cubic inch? _____

c. Cut out your net and tape it together to make a cubic inch.

d. How many cubes with an edge length of $\frac{1}{4}$ inch will fit inside your cubic inch? _____

e. Explain how you found the number of cubes with edge length $\frac{1}{4}$ inch that fit inside your cubic inch.

_____

 a. How many cubes with an edge length of $\frac{1}{2}$ inch will fit inside 1 cubic inch? _____

b. What is the volume of a cube with edge length of $\frac{1}{2}$ inch? _____

c. Write a number sentence to show how you found your answer to Part b.

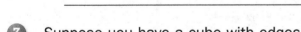

3 Suppose you have a cube with edges that measure $1\frac{1}{2}$ inches each.

a. How many cubes with an edge length of $\frac{1}{2}$ inch will fit inside your cube? _____

b. What is the volume of a cube with an edge length of $1\frac{1}{2}$ inches? _____

c. Write a number sentence to show how you found your answer to Part b.

_____

d. Explain how your answers to Problems 3a and 3b are related.

_____

_____

_____

_____

253

Comparing Volumes for Tissue Boxes

1 Below are diagrams that represent two tissue boxes sold at a local grocery store. Find the volume for each box to the nearest cubic inch.

Tissue Box A

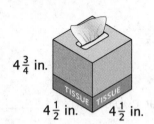

$4\frac{3}{4}$ in.

$4\frac{1}{2}$ in. $4\frac{1}{2}$ in.

Volume: _____

Price: $1.99

Tissue Box B

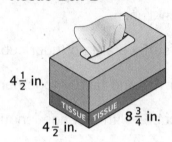

$4\frac{1}{2}$ in.

$4\frac{1}{2}$ in. $8\frac{3}{4}$ in.

Volume: _____

Price: $2.99

2 Which tissue box has the better price per cubic inch? _____

3 How many cubes with an edge length of $\frac{1}{4}$ inch would fit inside each tissue box?

Tissue Box A: _____   Tissue Box B: _____

4 Design a tissue box that would be reasonably priced at $4.99.
It should have a price per cubic inch that is similar to the boxes in Problem 1.

List the tissue box's dimensions:

Length: _____   Width: _____

Height: _____   Volume: _____

5 Explain how you decided on the dimensions for your tissue box in Problem 4.

_____

_____

_____

_____

Math Boxes

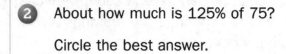

1 Write a value for the variable that makes the inequality true and one that makes it false.

	True	**False**
a. $g \geq -3$	_____	_____
b. $7 < t$	_____	_____
c. $\frac{2}{3} > w$	_____	_____
d. $f \leq 2.5$	_____	_____

SRB
210-211

2 About how much is 125% of 75?

Circle the best answer.

A. 70

B. 80

C. 90

D. 100

SRB
61-63

3 **a.** Plot and label points *A* and *B* on the coordinate grid.

A: $(-3, -3)$ 　　 B: $(-3, 2)$

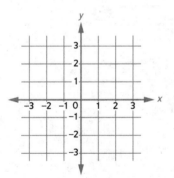

b. What is the distance between

points *A* and *B*? _____

SRB
96

4 Angel is wrapping a present for his mom. The present is in a box shaped like a cube that measures 13 cm on each edge. His sheet of wrapping paper is a rectangle that measures 26 cm by 52 cm, or 1,352 cm². Does he have enough wrapping paper to cover the cube? Explain.

_____

_____

_____

_____

SRB
263-264

5 **Writing/Reasoning** Explain how you used what you know about ratios to solve Problem 2.

Estimating the Volume of the Human Body

An average adult human male is about 69 inches (175 centimeters) tall and weighs about 170 pounds (77 kilograms). The drawings below show how a man's body can be approximated by 9 rectangular prisms.

The drawings use the scale 1 mm : 1 cm. This means that 1 millimeter in the drawing represents 1 centimeter of actual body length. The height of the drawing below is 175 millimeters. Therefore it represents a male who is 175 centimeters tall.

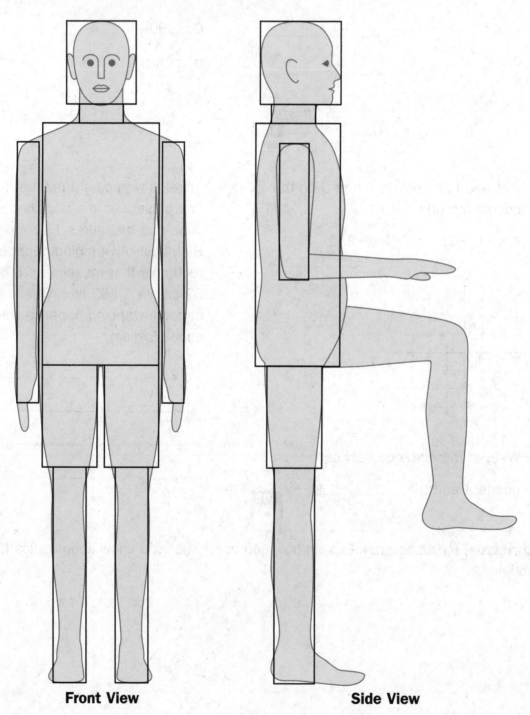

Front View   **Side View**

Estimating the Volume of the Human Body (continued)

1 **a.** Using a centimeter ruler, measure the dimensions of each rectangular prism shown on journal page 256. Use your measurements and the scale 1 mm : 1 cm to record an approximation for actual body dimensions in the table below. For example, if you measure the length of an arm as 72 millimeters, this will be recorded as 72 centimeters.

b. Calculate the volume of each body part and record it in the table. For the arm, upper leg, and lower leg, multiply the volume by 2.

c. Add the volumes of the parts to find the total volume of an average adult male's body. Your answer will be in cubic centimeters.

Body Part	Actual Body Dimensions (cm)			Volume (round to the nearest 1,000 cm³)
Head	Width: _____	Depth: _____	Height: _____	* 1 = _____
Neck	Width: _____	Depth: _____	Height: _____	* 1 = _____
Torso	Width: _____	Depth: _____	Height: _____	* 1 = _____
Arm	Width: _____	Depth: _____	Height: _____	* 2 = _____
Upper leg	Width: _____	Depth: _____	Height: _____	* 2 = _____
Lower leg	Width: _____	Depth: _____	Height: _____	* 2 = _____
			Total Volume: About	_____

2 **a.** One liter is equal to 1,000 cubic centimeters. Use this fact to complete the statement:

An average adult male's body has a volume of about _____ liters.

b. One liter of water has a mass of 1 kilogram (kg). Using this as a guide, is your estimate for the total volume of an adult male's body reasonable? _____

c. Use the information at the top of journal page 256 to justify your answer.

_____

_____

3 Make a reasonable estimate.
The volume of an average adult male's torso is about _____% the total body volume.

Finding Percents in a Survey

Come up with a four-question survey to give to your class. The questions should have "yes" or "no" answers. For example, you might ask about favorite musicians or favorite breakfast foods. Write the questions below. Survey your classmates and keep track of the answers.

Questions	Boys		Girls	
	Yes	No	Yes	No

1. What is the total number of students who answered each question? _____

2. Summarize the results of your survey using percents.

   _____

   _____

   _____

3. Who might be interested in the results of your survey?

   _____

4. Write one more question that you can answer with this survey.

   _____

5. Assume that the survey is representative of sixth graders—that is, assume that the survey results would be the same for any group of sixth graders. If 200 sixth graders were surveyed, what would you expect the results to be for your first question?

   _____

Try This

6. Ask a different group of people to answer your questions. Are the results the same? _____

   Why do you think that is? _____

Math Boxes

1 The lowest and highest temperatures ever recorded in California are −45°F (in 1937) and 134°F (in 1913). How many degrees separate these extreme temperatures?

Number model: _____

Solution: _____

SRB
92-93

2 **a.** Use mathematical language to describe the expression $(1.2 - 0.4)^2$.

_____

_____

b. Evaluate

_____ $= (1.2 - 0.4)^2$

SRB
200-202

3 Compare Box Plots A and B.

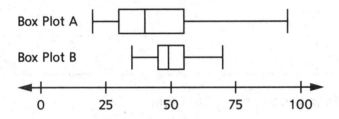

a. Which box plot has more variability? _____

b. Explain how you know. _____

_____

SRB
300-302

4 Estimate. Then calculate.

	Estimate	Answer
$\frac{3}{8} \div \frac{4}{5}$		
$\frac{4}{5} * 2\frac{1}{2}$		
$6\frac{2}{3} * 4\frac{4}{5}$		

SRB
189,
193-195

5 Four of John's 22 albums are classical music. Seven out of May's 32 albums are classical. Whose music collection has a higher ratio of classical albums? Support your answer.

_____

_____

_____

SRB
32

Comparing Linear, Square, and Cubic Measurements

Math Message

1 Work with a partner. Use grid paper.
One partner draws a net for a cube with edge lengths of 4 units and assembles it.
The other partner draws a net for a cube with edge lengths of 8 units and assembles it.
(You have to plan for this one so that it fits on the grid paper.)

2 Complete the table.

Edge Length (units)	Number Model for the Area of a Face (units^2)	Number Model for the Volume of the Cube (units^3)
2		
4		
8		
16		

3 Describe patterns and relationships that you see in the table.

_____

_____

_____

_____

4 Explain how the area and volume change when you double the edge length—for example, when you double the edge length from 2 units to 4 units.

_____

_____

Try This

5 What would happen if you tripled the edge length?

_____

_____

Why There Are No Giants

In the last lesson, you explored the volume of a human body.
An average adult human male is about 69 inches tall and weighs about 170 pounds.
Imagine a giant that is 10 times as tall as the average human male.

1 How many feet tall would the giant be?_____

2 A rule of thumb is that one
story of a building is about
10 feet tall. The giant would
be about as tall as building with _____
stories.

3 In a cross section of the
giant's thigh bone, the area
would be increased by a factor of _____.

4 The giant's volume (and
therefore the giant's weight)
would be increased by a factor of _____.

5 That means the giant's
bones (in relation to their
size) would have to support _____ times as
much weight.

6 Imagine an elephant. What do its legs look like?

_____

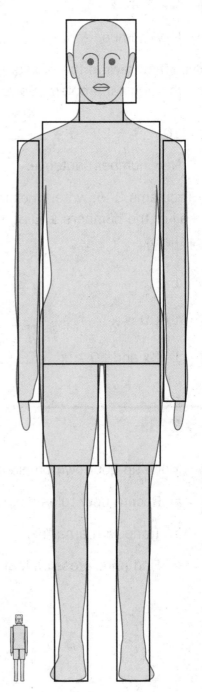

Try This

7 On a separate sheet of paper, sketch a picture of
what the giant might look like if its legs were big
enough to support its weight.

What do you have to consider to make this sketch?

_____

_____

_____

Solve. Then rewrite the problem, inserting parentheses so that the value of the expression is greater than the original problem. Solve your new problem.

1 $6 * 7 + 3^2 =$ _____

     New number sentence: _____

Solve. Then rewrite the problem, inserting parentheses so that the value of the expression is less than the original problem. Solve your new problem.

2 $10 - 8 \div 2 + 6 =$ _____

     New number sentence: _____

For Problems 3–5, write an expression for x to make each inequality true. For each expression, use all of the numbers 1–6 at least once, an exponent, and at least two different arithmetic operations.

3 $1 \geq x$ _____

4 $1{,}000 \leq x$ _____

5 $1 \leq x$ and $10 \geq x$ _____

Try This

6   Make up your own problem.

     a. Record your rules and your inequality.

     b. Trade with a partner.

     c. Find an expression that uses your partner's rules and makes the inequality true.

Math Boxes

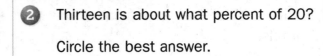

① Write a value for the variable that makes the inequality true and one that makes it false.

	True	**False**
a. $m \geq 0$	_____	_____
b. $12 < q$	_____	_____
c. $-4 > c$	_____	_____
d. $b \leq 0.5$	_____	_____

SRB
210-211

② Thirteen is about what percent of 20?

Circle the best answer.

A. 12%

B. 55%

C. 60%

D. 75%

SRB
61-63

③ Plot and label points A, B, and C on the coordinate grid.

A: (1, 2) B: (1, −1) C: (1, −3)

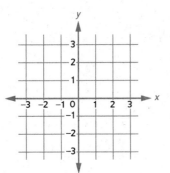

Write number sentences for the distances:

B to C _____

A to C _____

SRB
96

④ Lamar has a sheet of colored paper that measures 30 cm by 70 cm. He has three boxes that measure 10 cm by 8 cm by 7 cm each. Does he have enough paper to cover the surfaces of all three boxes? Explain.

_____

_____

_____

_____

SRB
263-264

⑤ **Writing/Reasoning** Based on what you know about x- and y-coordinates, explain how you know that the points in Problem 3 will be in a straight line.

263

Math Boxes
Preview for Unit 6

---

1 Write whether each statement is true or false.

a. $y > 12$, if $y = 9$ _____

b. $-2 \geq x$, if $x = 5$ _____

c. $m \leq 0$, if $m = -2$ _____

d. $z < \frac{3}{4}$, if $z = \frac{2}{3}$ _____

SRB
210-211

---

2 Write the next three numbers in each pattern.

a. 11, 23, 35, 47, _____, _____, _____

b. 7, 4, 1, −2, _____, _____, _____

c. 2, 5, 10, 17, _____, _____, _____

SRB
221

---

3 The prize for a contest is split into three cash prizes. The Purple Prize is x dollars. The Green Prize is twice as much as the Purple Prize. The Yellow Prize is $2 more than the Purple Prize.

Write algebraic expressions to show how much money each prize-winner gets.

Green: _____   Yellow: _____

If the total amount of prize money is $58, how much will each winner receive?

_____

_____

SRB
200-202

---

4 Evaluate.

a. $3g$, if $g = \frac{2}{5}$ _____

b. $5 - s$, if $s = 2$ _____

c. $2b - 1$, if $b = \frac{4}{3}$ _____

d. $h + \frac{3}{4}$, if $h = \frac{1}{4}$ _____

SRB
208-209

---

5 Which of the following are solutions to the equation $|x| = 4$?

Circle ALL that apply.

A. 4

B. $\frac{1}{4}$

C. 0

D. −4

SRB
208-209

---

6 Rewrite each number model using the Distributive Property.

a. $6 * (3 + \frac{1}{6}) =$ _____

b. $5 * (n + 5) =$ _____

c. $\frac{1}{2} * (t + 9) =$ _____

d. $1.5 * (k - 2) =$ _____

SRB
204-205

Substitution Number Puzzles

Math Message

1 Which number makes the equation below true: 1, 2, 3, or 4? _____

$$11 - x = 7 + x$$

Explain how you found the answer.

_____

_____

_____

_____

2 Each variable in the equations represents a different whole number between 1 and 10.

- A variable stands for the same number every time it appears.

- No two variables have the same value.

For each variable, tell what number will make the number sentences true.

Hint: Trial and error may be helpful.

$d - b = c$	$a =$ _____
$b + b = e$	$b =$ _____
$a * d = d$	$c =$ _____
$a + d = e$	$d =$ _____
$b - a = c$	$e =$ _____

3 Explain how you found your answers.

_____

_____

_____

_____

_____

_____

Using Trial and Error

If you substitute a number for a variable in an equation and the result is a true number sentence, then that number is a solution to the equation. One way to solve an equation is to try several test numbers until you find a solution. Each test number can help you close in on an exact solution.

This trial-and-error method may not result in an exact solution, but you can come quite close to an exact solution.

Use trial and error to find a solution for $\frac{1}{x} + x = 4$. If you cannot find an exact solution, find a number that is very close to the solution. This table shows the results of substituting several test numbers for x:

x	$\frac{1}{x}$	$\frac{1}{x} + x$	Compare ($\frac{1}{x} + x$) to 4
1	1	2	< 4
2	0.5	2.5	< 4
3	0.333 . . .	3.333 . . .	< 4
4	0.25	4.25	> 4

Based on the results above, it makes sense to try numbers between 3 and 4.

Results are rounded to the nearest thousandth.

x	$\frac{1}{x}$	$\frac{1}{x} + x$	Compare ($\frac{1}{x} + x$) to 4
3.9	0.256	4.156	> 4
3.6	0.278	3.878	< 4

Keep going. Get as close as you can to the exact answer.

Round your results to nearest thousandth.

x	$\frac{1}{x}$	$\frac{1}{x} + x$	Compare ($\frac{1}{x} + x$) to 4

Using Trial and Error (continued)

For each equation, try to get as close as possible to the exact solution.
Use the suggested test numbers to get started. Round numbers to the nearest thousandth.

① Equation: $2y * y = 47$

y	2y	2y * y	Compare (2y * y) to 47
1	2	2	< 47
5	10	50	> 47

My closest number: _____

② Equation: $z^2 - 5z = 30$

z	z^2	5z	$z^2 - 5z$	Compare ($z^2 - 5z$) to 30
6	36	30	6	< 30
8	64	40	24	< 30
9	81	45	36	> 30

My closest number: _____

Math Boxes

1 Use a tape diagram to solve the problem.

The ratio of bagels to croissants sold by a bakery is 4 : 5. The bakery sold a total of 1,800 bagels and croissants in one week.

How many bagels did they sell?

Solution: _____

SRB
45-48

2 Adela wears a pedometer. On Monday she walked $2\frac{3}{5}$ miles. On Tuesday she walked $1\frac{3}{4}$ miles, and on Wednesday she walked $2\frac{7}{10}$ miles. How many miles has she walked so far this week?

Number model: _____

Solution: _____

SRB
32

3 This data represents the number of books checked out of the school library by five students:

10, 15, 5, 10, 20

Find the mean absolute deviation.

m.a.d. _____

SRB
293-294

4 Calculate.

a. $1 + \frac{1}{2} \div 3 =$ _____

b. _____ $= \frac{1}{4} * (16 - 7)$

c. $3 \div \frac{2}{3} + 5 =$ _____

d. _____ $= 5 * 1\frac{1}{3} + 6 * \frac{2}{3}$

SRB
179-184,
203

5 **Writing/Reasoning** Explain how you drew and used a tape diagram to model and solve Problem 1.

Using Solution-Set Notation

Math Message

1 How many solutions are there for each equation? Circle the best answer.
Then list up to three solutions.

a. $3 + m = 3$ One   None   More than one   Solution(s): _____

b. $s + 8 = s$ One   None   More than one   Solution(s): _____

c. $w + 2 = 2 + w$ One   None   More than one   Solution(s): _____

d. $|e| + 3 = 10$ One   None   More than one   Solution(s): _____

---

2 Draw a line to match each equation with its solution set.
You may use an answer choice more than once or not at all.

$0 * x = 0$ **A.** {All numbers}

$x + 2 = 12$ **B.** { } or ∅

$x + 2 = x$ **C.** {10}

$x * 7 = 7 * x$ **D.** {2}

$\frac{8}{x} = 4$ **E.** {−10, 10}

$|x| = 10$

3 Record the solution sets for the equations below.

a. $\frac{1}{2}g * 2 = g$ Solution set: _____

b. $|k| = 12$ Solution set: _____

c. $j + 5 = j - 5$ Solution set: _____

4 Rewrite each equation as an inequality. Record the solution sets.

a. $0 * x = 0$ Inequality: _____   Solution set: _____

b. $x + 2 = x$ Inequality: _____   Solution set: _____

c. $x * 7 = 7 * x$ Inequality: _____   Solution set: _____

d. $\frac{8}{x} = 4$ Inequality: _____   Solution set: _____

Using Solution-Set Notation

(continued)

Set notation is useful for recording the solutions to inequalities.

5 **a.** The solution set is {all numbers less than 4}. Circle inequalities with this solution set.

$h > 4$ $h < 4$ $4 < h$ $4 > h$

b. Describe a situation one of the circled inequalities could represent.

_____

6 **a.** The solution set is {all numbers greater than 5}. Circle inequalities with this solution set.

$z + 8 < 10$ $6 < z + 1$ $6 > 5 + z$ $8 > 5z$

b. Explain how you found your answer for Problem 6a.

_____

_____

7 **a.** The solution set is { }. Circle inequalities with this solution set.

$b + 6 < b + 2$ $3b < b + 5$ $6 > 5 + b$ $8 > 3b$

b. Explain how you found your answer for Problem 7a.

_____

_____

8 Describe a solution set and write an inequality that has that solution set.

a. Solution set _____

b. Inequality _____

c. Describe a situation that your inequality could represent.

_____

_____

Try This

9 Write an equation or inequality for each solution set.

a. { } or $\varnothing$ _____   **b.** {0} _____

Plotting Points and Finding the Perimeter

① Roll two six-sided dice. Use the numbers shown on the dice and their opposites to generate four points. If you roll a and b, the points are (a, b); $(a, -b)$; $(-a, -b)$; and $(-a, b)$.

Record the numbers from your dice roll: _____

Points: M: (_____, _____)   N: (_____, _____)

          O: (_____, _____)   P: (_____, _____)

② Plot your points on the grid. Label your points: M, N, O, and P.
Connect your points to make Rectangle MNOP.

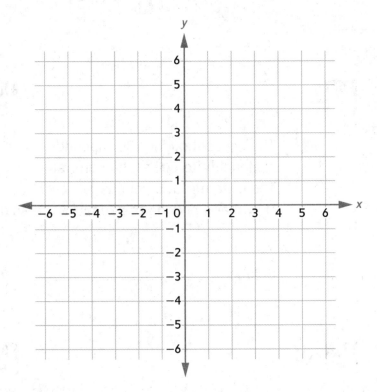

③ Find the perimeter of Rectangle MNOP. _____

Try This

④ Choose two connected points. Change the coordinates of both so that the line segment is 2.5 units longer (or 2.5 units shorter).

Plot the new points and label them K and L. Connect the new points.

List the coordinates and the line segment length.

K: (_____, _____)   L: (_____, _____)   Length: _____

271

Math Boxes

Math Boxes

---

1 Lamar was packaging his 84 model cars and 48 model trucks together in boxes to sell at his mom's garage sale.

He needed to make as many boxes as possible. He wanted to distribute all the cars and trucks equally among the boxes.

How many cars and trucks went into each box?

_____

How many boxes did he have for sale?

_____

SRB
32

---

2 Which of the following expressions is equivalent to $5(2x + 3)$?

Circle ALL that apply.

A. $10x + 3$

B. $10x + 15$

C. $8x + 2x + 15$

D. $25x$

SRB
206

---

3 Solve.

a. 10% of 180 _____

b. 25% of 82 _____

c. 30% of 15 _____

d. 12% of 250 _____

SRB
59-60

---

4 The school bought 130.5 pounds of peanuts to sell at the carnival. There were 0.375 pounds left over. How many pounds of peanuts were sold?

Solution: _____

SRB
32

---

5 Every year, the school uses bubble mixture to make giant bubbles at the carnival. This year the school used $1\frac{1}{2}$ times as much bubble mixture as last year. Last year the school used $4\frac{2}{5}$ gallons. How many gallons of bubble mixture did the school use this year?

Solution: _____

SRB
32

---

6 Write a statistical question that could be answered using the box plots on page 302 of the *Student Reference Book*.

_____

_____

_____

SRB
280

Solving Equations with Bar Models

Math Message

Bar-model diagrams are useful tools for solving equations with variables on both sides of the equal sign.

Study the example below, and then use a bar model to solve for *d* in Problem 1.

Example: $3v + 2 = 6 + v$ $v = 2$

3v	2
v	6

v	v	v	2
v	4		2

v	v	v	2
v	2	2	2

1 $3d + 12 = 20 + d$ $d = $ _____

2 Explain how you can check your answer to Problem 1.

_____

_____

Use a bar model to solve each of the following equations.

3 $6f + 7 = 4f + 11$ Solution: _____

4 $86 = 14 + 3b$ Solution: _____

5 $98 + x = 3x + 14$ Solution: _____

Solving Number Stories with Bar Models

Solve the number stories below. Use the space at the right for your bar-model diagrams.

1 Edward is 4 years more than three times as old as his sister Eileen. The sum of their ages is 20. Let *a* be Eileen's age.

Write an expression for Edward's age. _____

Write an equation to represent the situation.

_____

How old are Eileen and Edward?

Eileen: _____   Edward: _____

2 Lauren planted 15 more than twice as many tulips as roses in her yard. Let *c* be the number of roses.

Write an expression to represent the number of tulips.

_____

Lauren planted a total of 57 flowers.
Write an equation to model the situation.

_____

How many of each flower did she plant?

Number of roses: _____   Number of tulips: _____

3 Amy has cats, dogs, and fish. She has twice as many cats as dogs. She has 5 times as many fish as dogs.

Let *t* be the number of dogs. Write expressions to represent the number of cats and fish she has.

Cats: _____   Fish: _____

Amy has a total of 16 pets.
Write an equation to represent the situation.

_____

How many of each type of pet does she have?

Dogs: _____   Cats: _____   Fish: _____

274

Math Boxes

1 A chef cooks rice using a 1 : 2 ratio of dry rice to water. Her recipe calls for 9.5 cups of dry rice.

How much water does she need?

Solution: _____

SRB
32

2 Sasha's grandfather bought $4\frac{2}{3}$ pounds of apples. He used $3\frac{1}{4}$ pounds of apples to make applesauce. How many pounds of apples does he have left?

Number model: _____

Solution: _____

SRB
32

3 The numbers below represent minutes spent watching TV on a Friday night for six different students.

75, 30, 45, 120, 60, 75

Find the mean absolute deviation.

m.a.d. _____

SRB
293-294

4 Solve.

a. $9 * (\frac{1}{3} + 4) =$ _____

b. $8 + \frac{3}{5} * 10 =$ _____

c. _____ $= \frac{3}{4} * (6 + 7)$

d. _____ $= \frac{4}{3} + \frac{5}{6} \div \frac{1}{6}$

SRB
179-184,
203

5 **Writing/Reasoning** What do you know about the Distributive Property that could help you solve Problem 4c?

Math Boxes

275

Pan-Balance Problems

Math Message

 1   **a.** Let *m* be the number of cylinders
and *n* be the number of spheres.

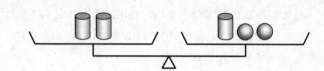

Write an equation that shows that
the weights of the two sides are equal. _____

b. Represent your equation with a bar model.

---

2   Use the pan balance in Problem 1.

weighs as much as _____ spheres.

Solve these pan-balance problems. In each figure, the two pans are in perfect balance.

3   One cube weighs

as much as _____ marbles.

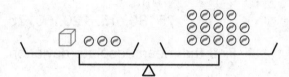

4   One cube weighs

as much as _____ oranges.

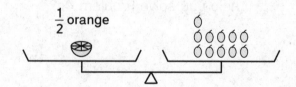

5   One whole orange weighs

as much as _____ grapes.

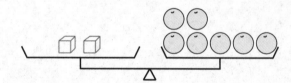

6   One block weighs

as much as _____ marbles.

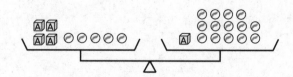

Pan-Balance Problems (continued)

7 One ☐ weighs

as much as _____ △s.

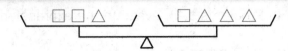

8 One ☐ weighs

as much as _____ marbles.

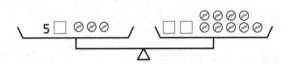

9 One block weighs

as much as _____ balls.

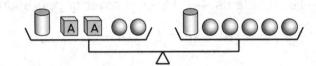

Try This

10 An empty bottle weighs as much as 6 marbles.

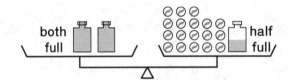

a. The content within a full bottle weighs

as much as _____ marbles.

b. A full bottle together with its content weighs

as much as _____ marbles.

c. Explain your solutions.

_____

_____

_____

_____

_____

_____

_____

_____

277

Order of Operations

In Problems 1–3, evaluate the expression. Then tell what you did first.

Expression   **First Step**

1 $17 - 3 * 5 =$ _____   _____

2 $26 - 7 * 3\frac{2}{5} =$ _____   _____

3 $12 - (\frac{5}{8} - \frac{2}{32}) =$ _____   _____

For Problems 4–7, use the order of operations to decide whether each equation is true or false.

4 $5 * (14 - 9) = 5 * 14 - 9$ _____

5 $1 + 18 \div \frac{1}{2} = 18 * 2 + 1$ _____

6 $\frac{6}{9} - 2 * 3 = \frac{6}{9} - 6$ _____

7 $3 * (4 - \frac{1}{7}) = 3 * 4 - 3 * \frac{1}{7}$ _____

8 Write an expression for each of the following numbers using five 5s. All values can be found using some combination of addition, subtraction, multiplication, and division.

Use the order of operations to write your expressions.

$1 =$ _____

$2 =$ _____

$3 =$ _____

$4 =$ _____

$5 =$ _____

$6 =$ _____

$7 =$ _____

Math Boxes

1 It is October 1st, and the volleyball and basketball teams are both practicing. The volleyball team practices every three days. The basketball team practices every five days.

When is the next day both teams will practice?

Solution: _____

SRB
32

2 Which expressions are equivalent to $9y + 63$?

Circle ALL that apply.

A. $9(y + 63)$

B. $3(3y + 6)$

C. $9(y + 7)$

D. $3(3y + 21)$

SRB
206

3 Solve.

a. 10% of _____ is 9.

b. 25% of _____ is 17.

c. 75% of _____ is 360.

d. 5% of _____ is 92.

SRB
59-6

4 The school bought 12.5 pounds of apples, 8.6 pounds of pineapples, and 11.35 pounds of grapes for a fruit salad.

How much fruit did the school purchase for fruit salad?

Solution: _____

SRB
32

5 What is the area of a rectangular rug that measures $8\frac{1}{4}$ feet by $10\frac{3}{4}$ feet?

Area: _____

SRB
253

6 Write a statistical question that could be answered with the graph.

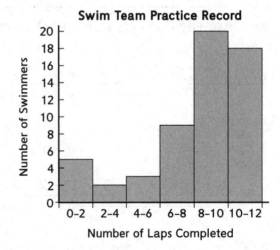

Swim Team Practice Record

_____

_____

SRB
280

279

A Balancing Act with Pan Balances

Math Message

1 These two pan balances are each in perfect balance.

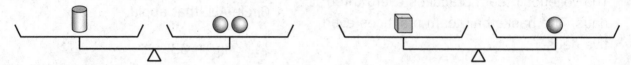

a. Use the relationships in the pan balances shown above to determine which of the pan balances below are balanced. Circle the ones that are in balance.

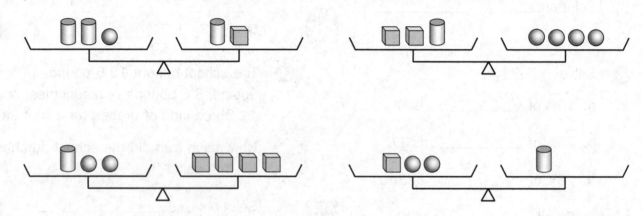

b. For any pan balance above that you did not circle, add or cross out objects to balance the pans.

---

2 **a.** Circle each pan balance that is balanced with equivalent expressions.

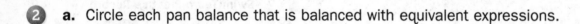

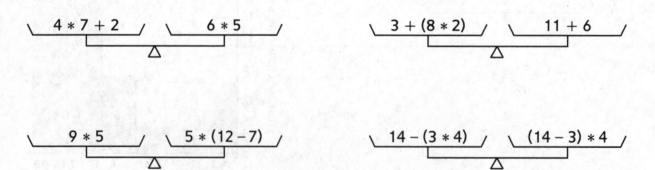

| 4 * 7 + 2 | 6 * 5 |
| 3 + (8 * 2) | 11 + 6 |

| 9 * 5 | 5 * (12 − 7) |
| 14 − (3 * 4) | (14 − 3) * 4 |

b. For any pan balance above that you did not circle, cross out one of the expressions and write a new one above it that will balance the pans.

280

3   Balance each pan balance. Record equivalent expressions in each pair of pans. Use expressions from the list below.

You will not use all of the expressions.

7 * 5 + 10

100 − 1

$\frac{1}{2}$ * (22 * 9)

9 * 5

22.5 * 2

2^4 − (7.5 * 2)

100 ÷ (25 * 4)

90 ÷ 2

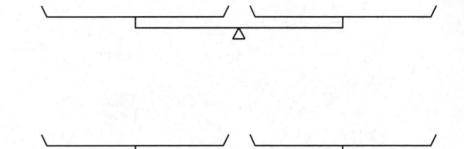

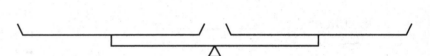

4   Find the value of the missing number that will balance each set of pans below. The same number is missing from both sides of a pan balance.

a.   □ + 3   2 * □

□ = _____

b.   2 * □ + 7   □ + 17

□ = _____

c.   □ * 20   80 ÷ □

□ = _____

d.   63 − □   2 * □ + 3

□ = _____

Math Boxes

Math Boxes

1 A small theater can hold as many as 100 people. Write an inequality to represent the situation.

Define a variable:

_____

_____

Inequality: _____

SRB
210-211

2 Find the area of the parallelogram.

5 cm

8.5 cm

Area: _____

SRB
254

3 Cindy ran 5 km in 26.5 minutes. Deirdre ran 15 km in 78 minutes. Whose average rate was faster?

Solution: _____

SRB
32

4 Volunteers at the county fair will give away snack packs. They bought 140 boxes of raisins and 350 small bags of almonds.

How many identical snack packs can they make using all the raisins and almonds? _____

How many boxes of raisins will be in each pack? _____

How many bags of almonds? _____

SRB
32

5 **Writing/Reasoning** Explain two ways you can determine who was faster in Problem 3.

_____

282

Combining Like Terms

Math Message

1   Complete each name-collection box by writing at least four equivalent expressions in the box.

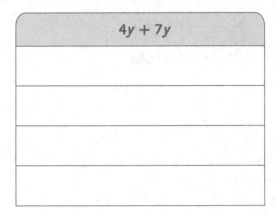

4y + 7y

7y − 4y

Algebraic expressions contain **terms.** The expression $4y + 7y$ is composed of two terms: $4y$ and $7y$. These are called **like terms** because they are multiples of the same variable.

To **combine like terms** means to rewrite the sum or difference of like terms as a single term.

To **simplify an expression** means to write the expression in a simpler form. Combining like terms is one way to do that.

2   Use the Distributive Property to combine the like terms.

Example: $4y + 7y$
$$= (4 + 7) * y$$
$$= 11y$$

a. $6y + 13y$

b. $12g - g$

3   Simplify each expression by combining the like terms.

a. $2z + 8z$ _____

b. $7w - w$ _____

c. $16.85f - 1.3f$ _____

d. $27k + 10.7k$ _____

e. $7q + 1\frac{3}{8}q$ _____

f. $12.8g - 8.2g$ _____

g. $15b - 5\frac{2}{3}b$ _____

h. $3n + 3n + 5n$ _____

283

Combining Like Terms (continued)

4 Use the Distributive Property to simplify these expressions.

Example: $6m + 10 + 28m + 9$
$$= (6 + 28) * m + (10 + 9)$$
$$= 34m + 19$$

a. $87 + 14t + 8t$

b. $38 + 105x + 74 + 68x$

c. $502 + 102u - 8u - 32$

d. $312 + 15r - 106 + 15r - 35$

5 Use the Distributive Property and combine like terms to simplify expressions with parentheses and multiplication.

Example: $10(x + 4) - 3x$
$$= 10x + 40 - 3x$$
$$= 40 + 10x - 3x$$
$$= 40 + (10 - 3)x$$
$$= 40 + 7x$$

a. $11(n + 5) + 105$

b. $25 + 8(b - 2)$

c. $42(z + 3) + 12(z - 4)$

Comparing Tables, Box Plots, and M.A.D.

The tables show the ten movies that took in the most money in 2012 and 2013.
Sales are in dollars rounded to the nearest million.

2012 Top-Grossing Movies

Movie	Gross
Marvel's The Avengers	623
The Dark Knight Rises	448
The Hunger Games	408
Skyfall	304
The Hobbit: An Unexpected Journey	303
Twilight Saga: Breaking Dawn Part 2	292
The Amazing Spider-Man	262
Brave	237
Ted	219
Madagascar 3: Europe's Most Wanted	216

2013 Top-Grossing Movies

Movie	Gross
The Hunger Games: Catching Fire	425
Iron Man 3	409
Frozen	400
Despicable Me 2	368
Man of Steel	291
Gravity	274
Monsters University	268
The Hobbit: The Desolation of Smaug	258
Fast & Furious 6	239
Oz the Great and Powerful	235

1. Which year do you think had greater variability in movie earnings? Explain.

_____

2. Make a box plot for each year.

2012 Top-Grossing Movies   **2013 Top-Grossing Movies**

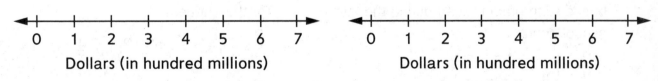

0   1   2   3   4   5   6   7   0   1   2   3   4   5   6   7
Dollars (in hundred millions)   Dollars (in hundred millions)

3. Use your box plots to explain which year you now think had greater variability.

_____

_____

4. The mean value rounded to the nearest million is $331 for 2012 and $317 for 2013.
   Find the mean absolute deviation (to the nearest million) for each year.

   2012: _____   2013: _____

285

1 Allison has a fish tank that is 20 in. by 10 in. by 12 in. What is the volume in cubic inches?

Solution: _____

The fish tank holds 10 gallons of water. Approximately how many cubic inches does 1 gallon of water occupy?

Solution: _____

SRB
259-261

2 Use the given ratio to find equivalent measurements.

1 meter : 100 centimeters

a. 3 m : _____

b. _____ : 700 cm

c. 4.5 m : _____

d. _____ : 9,500 cm

SRB
68-69

3 Write an inequality with the solution set {all numbers greater than 3}.

Inequality: _____

SRB
210-211

4 Write the next three numbers in each pattern:

a. 3, 7, 11, 15, _____, _____, _____

b. 1, 2, 4, 8, 16, _____, _____, _____

c. 5, 11, 17, 23, _____, _____, _____

SRB
221

5 Below is a box plot for the number of books sixth graders read during the last school year.

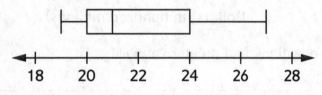

18   20   22   24   26   28

Find the IQR. _____

SRB
292-293

6 Fill in the circle next to the best estimate for the quotient.

$$153.12 \div 2.4$$

Ⓐ 77

Ⓑ 300

Ⓒ 7

Ⓓ 7.7

SRB
113-114

Identifying Equivalent Expressions

Math Message

1 Cross out expressions in the name-collection box that are NOT equivalent. Record one equivalent expression that could be added to the name-collection box.

$9(x + 1) + 3(x + 1)$	
$9x + 9 + 3x + 3$	$12x + 10 + 4 \div 2$
$7(x + 2) + 3$	$12x + 2$
$9x + 3x + 9 + 3$	$9x + 1 + 3x + 1$
$(30 - 18) + 10x + 2$	$12(x + 1)$

_____

2 Explain how you determined whether an expression was equivalent to $9(x + 1) + 3(x + 1)$.

_____

_____

_____

3 Cyrus says $9(x + 1) + 3(x + 1)$ and $7(x + 2) + 3$ are equivalent expressions. He substituted 1 for x and found that they were both equal to 24. How can they both equal 24 and still not be equivalent expressions?

_____

_____

_____

For Problems 4–5, find the simplest form for each expression.

4 $7 + (5 - 3) * x + 1 + x$ Simplest form: _____

5 $x + 2.5x + b + 5x$ Simplest form: _____

6 Simplify the expressions below. Are they equivalent? _____

$2x + 4 * 3 + 3x + 8$ $4x + x + x + 20$

Explain. _____

_____

Finding Equivalent Equations

Equations are equivalent when the corresponding expressions in the equations can be reduced to the same simplified forms. In addition, equivalent expressions have the same solution set.

1 Juno created a set of equivalent equations by generating equivalent expressions. For each new equation, describe what Juno did.

Original
Equation: $4y + 6 + 4 = 8(1 + y)$

Step 1: $4y + 10 = 8(1 + y)$ _____

Step 2: $4y + 10 = 8 + 8y$ _____

In Problems 2–5, combine like terms to find an equivalent equation that is simpler than the original.

2 $5h + 13h = 20 - 2$ Equivalent equation: _____

3 $2 + x + 2x + 4 = 2 * 16 + 10$ Equivalent equation: _____

4 $3(y + 2) = 4(y + 3) - 8$ Equivalent equation: _____

5 $5(z + 3) - 2.5z = 35 + z + 10$ Equivalent equation: _____

6 In Problems 2–5, solve the equations by finding the value of the variable.
Bar models or pan balances may help. Use the space below to show your work.

$h =$ _____ $x =$ _____ $y =$ _____ $z =$ _____

7 Explain why it might be easier to use the simplified equations you wrote for Problems 2–5 to find the solutions in Problem 6.

_____

_____

Math Boxes

1 Define a variable and write an inequality to represent the situation.

Buying a gift for no more than $30
Define a variable:

_____

Inequality: _____

Predicting at least 4 inches of snow
Define a variable:

_____

Inequality: _____

SRB
210-211

2 Find the area of the triangle.

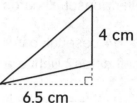

4 cm

6.5 cm

Area: _____

SRB
255-256

3 Car A goes 378 miles on 9 gallons of gas.

Car B goes 418 miles on 11 gallons of gas.

Which car gets better gas mileage?

Solution: _____

SRB
32

4 The clock in the living room chimes every 15 minutes.

The clock in the kitchen chimes every 20 minutes.

Both clocks chimed at 2:00 P.M. At what time will they next chime together?

_____

SRB
32

5 **Writing/Reasoning** Malia says the answer to Problem 4 is 7:00 P.M. because you multiply 15 by 20 to get 300 minutes. Explain how you know Malia did not find the next time the clocks will chime together.

289

The Same Number of Cards

Math Message

Two friends collect baseball cards.

1 Travis has 64 baseball cards and buys 3 new cards every week. When will Travis have 73 baseball cards? Define a variable and write an equation for Travis's situation.

Variable: _____

Equation: _____

Use a bar model to solve your equation.

How many more weeks will it take Travis to reach a total of 73 baseball cards? _____

2 Robert has 40 baseball cards and buys 7 new cards every week.

Construct a table to figure out when Travis and Robert will have the same number of baseball cards.

How many more weeks will it take for Travis
and Robert to have the same number of cards? _____

Math Boxes
Preview for Unit 7

Math Boxes

---

1 Use words to write what each algebraic expression represents.

a. $6 + y$ _____

b. $5j$ _____

c. $\frac{h}{8}$ _____

d. $4(x + 7)$ _____

SRB
200-202

---

2 The price for 8 tickets to the school Fun Fair is \$42. Find a unit rate to help you complete the table.
(Each ticket costs the same amount.)

Number of Tickets	Cost
1	
2	
10	
x	

SRB
43-44,
49-50

---

3 Graph the solution set for $j \leq 2$ on the number line below.

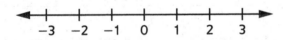

SRB
210-211

---

4 Substitute the given number for the variable. Record whether the resulting number sentence is true or false.

a. $5x = 75$, $x = 15$ _____

b. $x - \frac{3}{5} = 2\frac{1}{4}$, $x = 3$ _____

c. $2x - 4.2 = x + 1.7$, $x = 5.9$ _____

SRB
207-208

---

5 Plot and label the points.
Connect the points with line segments and plot the fourth vertex to make rectangle *ABCD*. Record the ordered pair for point *D*.

A: (−3, 2.5)

B: (−3, −2)

C: (2, −2)

D: (_____, _____)

What is the length of side *BC*? _____

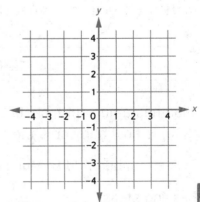

SRB
265-267

Number Tricks

Math Message

1 Abraham enjoys number tricks. Below is one of his favorites:
Pick a secret number. Perform each step and record the resulting value.

    a. Add 3 to the number.   _____

    b. Multiply the sum by 10.   _____

    c. Then divide by 5.   _____

    d. Subtract 6.   _____

    e. Divide by 2.   _____

    What number did you get?   _____

2 How does Abraham's number trick work?

_____

_____

_____

3 Marcy took her secret number, added 5, doubled the sum, and subtracted 2.
Her result was 58.

    a. What was Marcy's secret number? _____

    b. What steps did you complete to find Marcy's secret number?

    _____

    c. How did you know what steps to complete and in what order to complete them?

    _____

    _____

4 Juan has a number trick, but he forgot the last two steps. Figure out the last two steps so that his final number is the same as his beginning number.

Pick a number. Add 8. Multiply by $\frac{1}{2}$.

Missing steps: _____

Using Inverse Operations to Solve Equations

Use inverse operations to solve these equations. Show each operation. Check your work.

1 $4,509 = x - 748$

2 $\frac{y}{7} = 18$

Check:

Check:

3 $5.75 = d + 0.29$

4 $1\frac{2}{3} = 5k$

Check:

Check:

5 $18 = 5s$

6 $3\frac{1}{8} = m - 2\frac{1}{4}$

Check:

Check:

Math Boxes

① Kori wants to cover her pencil box in decorative paper. The pencil box is a rectangular prism that measures 10 cm tall, 18 cm wide, and 25 cm long.

How much paper does she need?

Solution: _____

SRB
32

② Draw a line to estimate where the mean of the distribution is.

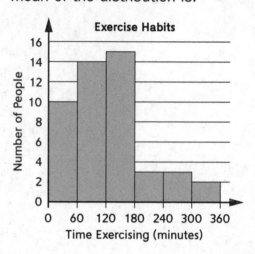

SRB
284-288

③ The pans are balanced. Complete the sentence for the balanced pans.

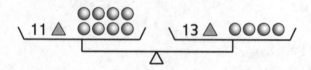

11 △ | 13 △

1 △ weighs as much as _____ marbles.

SRB
217-218

④ Alan is making spaghetti for the picnic at the end of a 40-mile bike ride. Each batch requires $3\frac{1}{2}$ jars of pasta sauce.

He plans to make $4\frac{1}{2}$ batches of spaghetti. How many jars of pasta sauce does he need?

Number model: _____

Solution: _____

SRB
32

⑤ **Writing/Reasoning** Explain how you could estimate to check your answer to Problem 4.

294

Equivalent Equations and Pan Balances

Math Message

1 Determine whether the two equations in each set are equivalent.
Be prepared to discuss how you decided whether they are equivalent or not.

Set A	**Set B**	**Set C**
$4x + 3x = 21$	$3(x + 2) = 7 + x$	$x = 5$
$7x = 21$	$3x + 2 = 7 + x$	$5x = 25$

Equivalent? _____   Equivalent? _____   Equivalent? _____

---

2 Start with the original pan-balance equation. Do the first operation on both sides of the pan balance and write the results on the second pan balance.
Complete the third and fourth pan balances in the same way.

Original pan-balance equation:

Operations (in words)

Multiply by 3.

Add 18.

Add 2x.

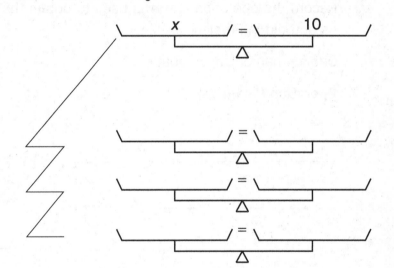

3 Now do the opposite of what you did in Problem 2. In words, record the operation you used to obtain the results on each pan balance.

_____

_____

_____

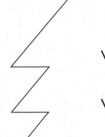

4 Record the results of the operation on each pan, as you did in Problem 2.

Original pan-balance equation:

Operation (in words)

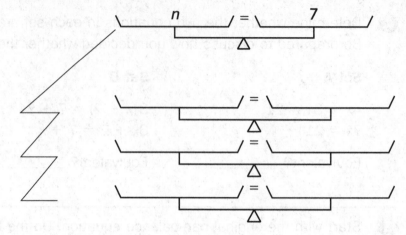

n = 7

Subtract 2.

=

Multiply by 4.

=

Add $2n$.

=

5 Check that 7 is a solution to each pan-balance equation in Problem 4.

6 Record the operation that was used to obtain the results on each pan balance, as you did in Problem 3.

Original pan-balance equation:

Operation (in words)

$4 * (n - 2) + 2n$ = $20 + 2n$

_____

$4 * (n - 2)$ = 20

_____

$n - 2$ = 5

_____

n = 7

7 Check that 7 is a solution to each pan-balance equation in Problem 6.

8 Original equation:

Operation (in words)

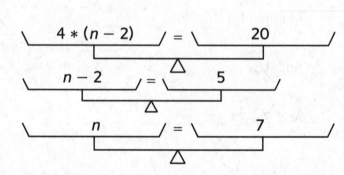

$3p + 6$ = 10

_____

$3p$ = 4

_____

p = $\frac{4}{3}$

9 Check that $\frac{4}{3}$ is a solution.

Building and Solving Equations

1. Use the riddle solution your teacher will give you. Do NOT show anyone the letters.
   Change each letter of the riddle solution into a number using the chart below.

A	B	C	D	E	F	G	H	I	J	K	L	M	N	O	P	Q	R	S	T	U	V	W	X	Y	Z
1	2	3	4	5	6	7	8	9	10	11	12	13	14	15	16	17	18	19	20	21	22	23	24	25	26

2. Build an equation for each letter by applying operations to both sides of the equation.
   Record the operations and write the new equivalent equations.

Check your work by substituting the original value of the variable in each equation.

Original equation: _____ w _____ = _____

Operation (in words)

_____   = _____

_____   = _____

_____   = _____

Original equation: _____ x _____ = _____

Operation (in words)

_____   = _____

_____   = _____

_____   = _____

Original equation: _____ y _____ = _____

Operation (in words)

_____   = _____

_____   = _____

_____   = _____

Original equation: _____ z _____ = _____

Operation (in words)

_____   = _____

_____   = _____

_____   = _____

3 When you have checked your equations, read the final equations you created to your partner.

4 Record and solve the equations your partner reads to you.

Equation for *w*: _____   Solution set: _____

Equation for *x*: _____   Solution set: _____

Equation for *y*: _____   Solution set: _____

Equation for *z*: _____   Solution set: _____

5 Use the table in Problem 1 to change the numbers in your solutions back to letters, and figure out the solution to one of the riddles below.

Record the answer next to its matching riddle.

Check with classmates for answers to the other three riddles.

Riddle 1: I am weightless, but you can see me.
　　　　　Put me in a bucket and I'll make it lighter. What am I?　　_____

Riddle 2: You can catch me but not throw me. What am I?　　_____

Riddle 3: Feed me and I will live. Water me and I will die. What am I?　　_____

Riddle 4: These four letters frighten a thief. What are they?　　_____

Rates and Ratios

Solve the ratio and rate problems below.

1 Complete the ratio/rate tables and use them to answer the questions.

   a. Fox Middle School surveyed students to learn how they travel to school and how their
   parents travel to work. The ratio of students taking a bus or walking to students being
   driven was 3 : 2. For parents commuting to work, it was 4 : 6.

   Students

Bus/Walking	3			
Car	2			

   Parents

Bus/Walking	4			
Car	6			

   b. Out of 30 students, how many would you
   expect to take a bus or walk to school?   _____

   Out of 30 parents, how many would you
   expect to take a bus or walk to work?   _____

   c. If 330 students attend Fox Middle School on a
   given day, how many would you expect to arrive by car?   _____

   If 330 parents go to work on a given day,
   how many would you expect to arrive by car?   _____

2 Ignacio's car traveled 362 miles on 10.4 gallons of gas. Joanne's car went 275 miles on
8.5 gallons. Which car got better gas mileage? Justify your answer.

_____

_____

3 The ratio of days Jo typically rides her bike to work to days she drives is 3 : 2. If she works
20 days this month, how many days would you expect her to bike to work? Explain.

_____

_____

Math Boxes

Math Boxes

1 What is the volume of a cube with a side length of 4.5 inches? Show your work.

Volume: _____

SRB 260-261

2 Use ratios to convert.

1 yard : 36 inches

a. 5 yd : _____ in.

b. _____ yd : 396 in.

c. 7.5 yd : _____ in.

d. 0.5 yd : _____ in.

SRB 68-69

3 Write an inequality with the solution set {all numbers less than or equal to −4}.

_____

SRB 210-211

4 Write the next three numbers in each pattern.

a. 12, 15, 18, 21, _____, _____, _____

b. 17, 32, 47, 62, _____, _____, _____

c. 192, 96, 48, 24, _____, _____, _____

SRB 221

5 Below is a box plot for bird body temperatures.

Bird Body Temperatures (°F)

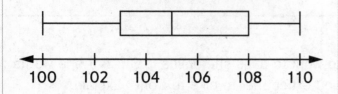

100   102   104   106   108   110

Find the IQR. _____

SRB 292-293

6 George bought 4.5 lb of pears at $1.90 per lb. How much money did he spend? Fill in the circle next to the best answer.

Ⓐ $2.60

Ⓑ $6.40

Ⓒ $8.55

Ⓓ $2.30

SRB 32

Solving Multistep Equations

Math Message

1 Model and solve the equation $4t + 6 = t + 15$ using a bar model or a pan-balance model.

Solve the equations in Problems 2–9. Use each of these strategies at least once:

Trial and Error   Bar Model   Pan-Balance Model   Inverse-Operations Strategy

Plan ahead so you are using a strategy or model that you think works best for a given equation.

2 $2x + 5 = 17$

3 $24 = 4a + 8$

Solution: _____

Solution: _____

4 $8b + 21 = 15b$

5 $12c - 5 = 9c + 22$

Solution: _____

Solution: _____

6 $\frac{2}{3}h + 2 = h$

7 $3d + 7.1 = 2.5d + 11.5$

Solution: _____

Solution: _____

8 $3f + f + 3 = 15$

9 $5(g - 2) = 3(g + 2)$

Solution: _____

Solution: _____

Math Boxes

1 The school dance has a math theme. Molly wants to put up decorated cubes. She has a rectangular piece of fabric that measures 20 inches by 22 inches. Her cubes have 8-inch edges. How many cubes can Molly cover with the fabric?

Solution: _____

SRB
32

2 On the graph for the food drive, draw a line to estimate where the mean of the distribution is.

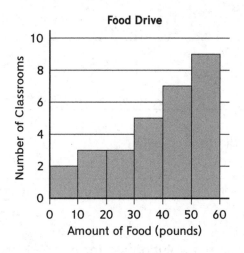

What does the mean tell you?

_____

_____

SRB
284-288

3 Solve the pan-balance problem.

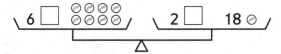

One ☐ weighs as much

as _____ marbles.

SRB
217-218

4 Flora is making salsa for a party. Every 1 cup of chopped tomatoes makes $2\frac{3}{4}$ cups of salsa.
She uses $1\frac{2}{3}$ cups of tomatoes.
How much salsa does she make?

Number model: _____

Solution: _____

SRB
32

5 **Writing/Reasoning** Explain how you decided where to place the line for the mean in Problem 2.

303

1 Write in words what each algebraic expression represents.

a. $y + 7$

_____

b. $5 - z$

_____

c. $8(x - 2)$

_____

SRB
200-202

2 The computer printer prints 30 pages in 2 minutes. Complete the table below.

Pages	Minutes
15	
45	
75	
	20
600	

SRB
43-44,
49-50

3 Graph the solution set for $m > -1$ on the number line below.

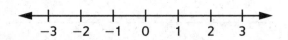

SRB
210-211

4 Substitute the number given to determine if it makes the equation true or false.

a. $6.2x = 8.37$, $x = 1.35$ _____

b. $\frac{x}{7} = 4.9$, $x = 34.1$ _____

c. $8x - 5.2 = 20$, $x = 3.15$ _____

d. $4x - 2 = x + 6.5$, $x = 2.8$ _____

SRB
207-208

5 Plot and label the points.

Connect the points with line segments and plot the fourth vertex to make parallelogram *CABD*.

A: $(-4, 2)$ B: $(1, 2)$

C: $(-2, -3)$ D: (_____, _____)

Record the ordered pair for the missing vertex at *D*.

What is the length of $\overline{AB}$? _____

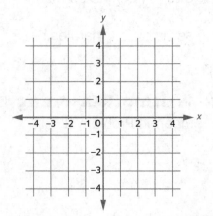

SRB
265-267

Mystery Numbers and Inequalities

Math Message

1 Harry and Susan are playing *Number Squeeze*, a game they played in earlier grades. Harry chooses a mystery number (*x*) between 0 and 50. Susan is guessing his number.

Susan guesses 7. Harry gives her the clue $x > 7$.

Susan guesses 45. Harry gives her the clue $x > 45$.

What number should Susan guess next? _____

Explain. _____

_____

_____

Theo and Margaret played a game of *Number Squeeze*.

For Problems 2–5, do the following:

- Represent each of the two *Number Squeeze* clues with an inequality.
  Describe the solution sets for the inequalities.

- Graph the solution set that makes both inequalities true.

- List three numbers that could be the mystery number.
  Check that they are in the solution sets for both inequalities.

2 Theo represents his mystery number with the variable *y*.

a.

Clue	The sum of *y* and 3 is less than 4.	The number *y* is greater than −4.
Inequality		
Solution Set		

b.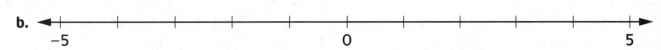
$$-5 \qquad\qquad\qquad 0 \qquad\qquad\qquad 5$$

c. Three numbers that *y* could be: _____

Mystery Numbers and Inequalities (continued)

3 Margaret represents her mystery number with the variable *p*.

a.

Clue	One-third of *p* is less than or equal to 1.4.	The number *p* is greater than 4.
Inequality		
Solution Set		

b.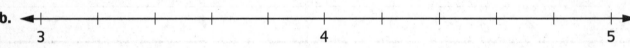

c. Three numbers that *p* could be: _____

4 Theo represents his mystery number with the variable *x*.

a.

Clue	The product of *x* and 3 is less than or equal to 27.	The number *x* is less than $8\frac{1}{2}$.
Inequality		
Solution Set		

b.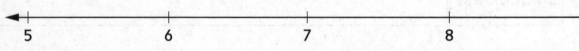

c. Three numbers that *x* could be: _____

Mystery Numbers and Inequalities (continued)

5 Margaret represents her mystery number with the variable *b*.

a.

Clue	When you subtract 7 from *b*, the difference is greater than 3.	The number *b* is less than or equal to 10.4.
Inequality		
Solution Set		

b.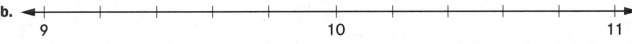

9   10   11

c. Three numbers that *b* could be: _____

6 **a.** Write two inequalities that could be clues for the following graph:

0   1   2   3   4   5   6

Inequality A: _____   Inequality B: _____

b. Write a different set of inequalities that could also represent the graph in Part a.

Inequality C: _____   Inequality D: _____

Try This

7 Margaret gives Theo two clues about her mystery number. Theo says that, based on her clues, the solution set is the null (or empty) set. Write two clues Margaret might have given.

_____

_____

_____

Why is { } the solution set? _____

_____

_____

307

Math Boxes

(1) Evaluate.

   a. $6^2 - 14 \div 7 =$ _____

   b. $15 \div \frac{1}{2} - 12 =$ _____

   c. _____ $= \frac{4}{5} * 7 - 2$

SRB
188, 193, 203

(2) Solve.

   a. 10% of n is 9; $n =$ _____

   b. 25% of x is 320; $x =$ _____

   c. 15% of y is 9; $y =$ _____

SRB
64

(3) Gauston Middle School has 267 students. Each cafeteria table holds 20 students. How many tables do they need to seat all of the students at the same time?

Solution: _____

SRB
32

(4) Write an algebraic expression to represent each situation.

   a. The perimeter of an equilateral triangle with sides of length b

      Expression: _____

   b. \$15 split equally among v friends

      Expression: _____

SRB
200-202

(5) Jenny runs 5 kilometers in 29.5 minutes. Jesse runs 8 kilometers in 44 minutes. Sarah runs 10 kilometers in 55 minutes. Who runs the fastest?

Circle the best answer.

   A.   Jenny

   B.   Jesse and Jenny

   C.   Sarah

   D.   Jesse and Sarah

SRB
49-50

(6) Leo bought 1.5 pounds of pears. Pears cost \$2.35 per pound. How much money did he spend?

Solution: _____

SRB
32

Math Boxes

Making Wise Salad Choices

Math Message

The Green Grocery has a salad bar.
The price of a salad is determined by its weight in pounds.
All items from the salad bar are also sold separately in the grocery store.
Below is the approximate price per pound for each item when sold separately.

Mesclun salad greens	$7.99	Celery	$0.99	Cheddar cheese	$6.99
Organic spinach	$6.67	Cucumbers	$2.15	Gorgonzola cheese	$17.99
Romaine lettuce	$3.06	Onion	$1.99	Almonds	$8.00
Red peppers	$2.97	Cherry tomatoes	$3.99	Walnuts	$9.99
Broccoli	$1.75	Sun-dried tomatoes	$6.67	Hard-boiled eggs	$1.67
Carrots	$1.69	Dried cranberries	$4.99	Bacon bits	$21.28

1. Archie and Natalia, the store managers, are trying to decide how much to charge per pound for their salad bar. Archie suggests charging at least $3.99. Define a variable and write an inequality to represent Archie's suggestion.

_____

Explain why you think Archie's price is reasonable, too high, or too low.

_____

_____

_____

After more discussion, Archie and Natalia decide to charge $7.99 per pound for the salad bar. For Problems 2–3, compare the costs. Represent each situation with an inequality.

Let x be the number of pounds. Record and graph the solution set to the nearest hundredth.

2. How many pounds of bacon bits sold separately could you buy for up to $6?

   a. Record the solution set for x using set notation.

   _____

   b. Record inequalities to represent the solution set. _____

   c. Graph the solution set for x that makes both inequalities true.

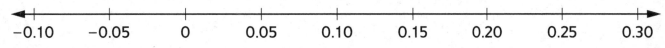

309

3 How many pounds of bacon bits from the salad bar could you buy for up to $6?

a. Record the solution set for *x* using set notation.

_____

b. Record inequalities to represent the solution set. _____

c. Graph the solution set for *x* that makes both inequalities true.

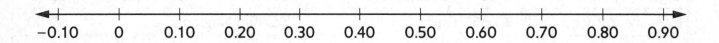

−0.10   0   0.10   0.20   0.30   0.40   0.50   0.60   0.70   0.80   0.90

4 The Green Grocery sells whole green peppers for $0.99 per pepper.
Let *z* be the number of green peppers.
What is the number of green peppers you might buy separately for up to $6?

a. Record the solution set for *z* using set notation.

_____

b. Record inequalities to represent the solution set. _____

c. Graph the solution set for *z* that makes both inequalities true.

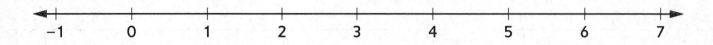

−1   0   1   2   3   4   5   6   7

5 **a.** Describe how your graph for Problem 4 is different from the graphs for Problems 2–3.

_____

_____

b. Explain why the graphs are different.

_____

_____

_____

_____

Creating Two Salads

1 Using the price information on journal page 309, create two salads that meet the following constraints:

- For Salad A: Cost at salad bar < Cost of buying items separately
- For Salad B: Cost at salad bar > Cost of buying items separately
- Let x be the total weight of the salad in pounds: $x \leq 0.6$ and $x > 0.3$.
- Let p be the number of ingredients: $p \geq 3$ and $p < 7$.
- Let t be the total cost of a salad: $t < \$15$ and $t > \$1$.

Salad A Ingredients	Quantity (pounds)	Grocery Store Cost ($)	Salad Bar Cost ($)
Total Cost			

Salad B Ingredients	Quantity (pounds)	Grocery Store Cost ($)	Salad Bar Cost ($)
Total Cost			

Finding the Best Ticket Prices

Next weekend Joe plans to see his favorite band, The Hawks, in concert.
Solve the problems. Round percents to the nearest 0.1%.

1   Joe goes online to buy a concert ticket for $80.
    When he clicks "BUY," an $18 service fee is added.

    What percent of the ticket price was added for the service fee?   _____

2   Joe only has $80. About what percent of the
    total concert ticket price of $98 can he afford to pay?   _____

3   Joe looks at other websites and finds a discount ticket company called Delilah's Deals.
    They advertise that their total price, including fees, is 90% of the original price of $80.

    How much would Joe pay for a ticket from Delilah's Deals?   _____

4   Joe knows that sales tax is 5.5%.

    How much will the ticket at Delilah's Deals cost after tax?   _____

5   Joe buys his ticket from Delilah's Deals. At the concert, Joe plans to buy a souvenir T-shirt
    for $25 and snacks for $10. Add sales tax of 5.5% to the purchases.

    If Joe had $80 before buying his ticket, how much more does he
    need to buy the T-shirt and snacks? (Make sure to add the tax.)   _____

Math Boxes

Math Boxes

---

① Two-thirds of Tyrone's books are mystery novels. Tyrone has 42 books.

How many mystery novels does he have?

Number model: _____

Solution: _____

SRB
32

---

② A cube has a side length of $\frac{1}{2}$ foot.

a. What is the cube's surface area?

Surface area: _____

b. What is the cube's volume?

Volume: _____

SRB
262-264

---

③ The vertices of triangle *ABC* are as follows:

    A: (0, 6)

    B: (0, −3)

    C: (−3, 4)

What is the length of side *AB*?

Solution: _____

SRB
96

---

④ Solve. Write the solution sets using set notation.

a. $4x = 28$

_____

b. $x - 3 = x + 3$

_____

c. $x + 5 = x + 4 + 1$

_____

SRB
208

---

⑤ **Writing/Reasoning** Explain how you can use what you know about absolute value to find the length of side *AB* in Problem 3.

Introducing Spreadsheets

Math Message

1 Complete the chart. Record an algebraic expression to represent each described operation.

Operation	Algebraic Expression
The sum of A and B	
The product of C and D	
8% of E	
The sum of F, G, H, I, and J	
The product of K and itself	

2 Use the spreadsheet below to answer the questions.

Boxes				⊠
D4	▼ ⊙	fx	= A4*C4	
	A	B	C	D
1		Budget for Book Fair		
2				
3	Quantity	Items	Unit Price ($)	Cost ($)
4	5	Rental Tables	9.25	46.25
5	25	Posters	3.15	78.75
6	4	Receipt books	2.72	10.88
7	100	Book covers	1.49	149
8	100	Bookmarks	0.26	26
9			Subtotal	310.88
10			7% tax	21.76
11			Total	332.64
12				

a. What is the title of the spreadsheet? _____

b. Which cell's formula is in the address box? _____

c. What formula is in cell D4? _____

d. What does the formula in cell D4 calculate? _____

e. What formula is in cell D5? _____

f. What formula is in cell D10? _____

g. What does the formula in D10 calculate? _____

h. If you changed the number of rental tables to 8, which cells would change? _____

Solving a "Squares" Problem

1 **a.** Complete the spreadsheet.

b. What formula will you enter in cell B4?

_____

c. What formula will you enter in cell C4?

_____

d. Use the data in the spreadsheet to make the graphs below. Connect the points.

☐ Boxes			☒
B3 ▼ ○ *fx*	= 4*A3		

	A	B	C	D
1		Squares		
2	Side (in.)	Perimeter (in.)	Area (in.²)	
3	1	4	1.00	
4	1.5			
5	2			
6	2.5			
7	3.5			
8	4			
9	5			
10	5.5			

Graph 1:

List points that you will graph: _____

_____

_____

_____

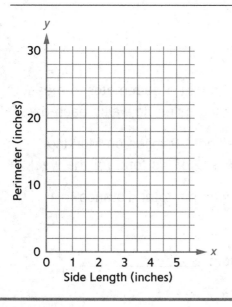

Graph 2:

List points that you will graph: _____

_____

_____

_____

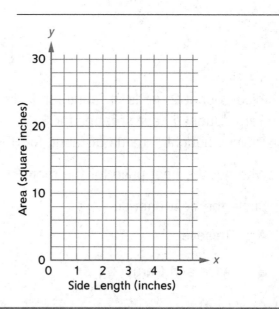

Try This

2 Use the graph to estimate the side length for a square with an area of 20 in.². _____

315

Math Boxes

1 Evaluate.

   a. $4^2 + 14 \div 7 =$ _____

   b. $25 \div \frac{1}{4} - 10 =$ _____

   c. _____ $= \frac{2}{3} * 6 - 2.5$

SRB
203

2 Solve.

   a. 20% of h is 140; $h =$ _____

   b. 5% of j is 320; $j =$ _____

   c. 30% of g is 15; $g =$ _____

SRB
64

3 Natural Bakery makes granola bars. Each granola bar weighs 4 ounces. Natural Bakery has 768 ounces of granola.

How many bars can they make?

Solution: _____

SRB
32

4 Write an algebraic expression to represent each statement.

   a. The sum of 2 times g and 17

      Expression: _____

   b. Three less than the product of 18 and h

      Expression: _____

SRB
200-202

5 Alice bought 2 yards of fabric for $14. Gary bought 7 feet of fabric for $13. Trevelle bought 9 yards of fabric for $56.

Who got the best price for the fabric?

Circle the best answer.

   A.   Trevelle

   B.   Alice and Gary

   C.   Gary

   D.   Alice and Trevelle

SRB
49-50

6 Jorge bought a pedometer to measure how many miles he walks each day.

He walked 1.2 miles on Monday. This was $\frac{1}{6}$ of his total for the week.

What was his total distance?

Solution: _____

SRB
32

316

Using Spreadsheets for Trial and Error

Math Message

1 Suppose you have a rectangle that is twice as long as it is wide. Let *a* be the width.

   a. What expression represents the length? _____

   b. What expression represents the perimeter? _____

   c. Use trial and error to find the width and the length if the perimeter is 42 m.

   Width: _____   Length: _____

A spreadsheet can be set up to use a trial-and-error strategy.

2 Describe how you could use the spreadsheet pictured here to help solve the Math Message problem.

Include a description of the formulas you would use and which cells would have formulas.

☐ Boxes				⊠
	A	**B**	**C**	**D**
1		Finding Perimeters		
2	**Width**	**Length**	**Perimeter**	
3	1			
4				
5				

_____

_____

3 Find the dimensions of rectangles that are similar to the Math Message rectangle (length is twice the width) but have different perimeters. Set up a spreadsheet to use trial and error.

Record the length and width for each perimeter.

Width	Length	Perimeter
		108 cm
		144 in.
		204 ft
		240 m

Try This

4 If given the perimeter, as in Problem 3, what formulas could you use to find the width and length when the length is twice the width.

_____

Using Spreadsheets
for Trial and Error (continued)

For Problems 5–7, solve each problem for the given value. Enter formulas and test various numbers until you find the solution. Use a spreadsheet program or your calculator. Record the formulas you used in the spreadsheets and your final answers.

5 Tomás, Richard, and Hannah collect baseball cards. Richard has 4 times as many cards as Tomás. Hannah has 10 more cards than Richard. The three collectors have 136 cards all together. Find the number of cards each collector has.

	A	B	C	D
	Tomás	**Richard**	**Hannah**	**Total**
1				
2	Test Number			

☐ Boxes

Answer: Tomás _____ Richard _____ Hannah _____

6 Four consecutive even numbers (like 10, 12, 14, and 16) are added together. Find the greatest four consecutive even numbers whose sum is less than 1,000.

	A	B	C	D	E
	1st Number	**2nd Number**	**3rd Number**	**4th Number**	**Sum**
1					
2	Test Number				

☐ Boxes

Answer: _____

7 Each of the two equal sides in an isosceles triangle has a length that is three times as long as the length of the third side. Find the lengths of the sides if the perimeter is 245 cm.

	A	B	C
	Length of Third Side	**Length of 1 Equal Side**	**Perimeter**
1			
2	Test Number		

☐ Boxes

Lengths of the three sides: _____

318

Comparing Game Results

1. Play four rounds of *Multiplication Top-It* in groups of four.

   At the end of each round, list the number of cards each player has in the table below.

Round 1	Round 2	Round 3	Round 4

2. Find the five-number summary for your group's sixteen data values (one value for each person for each round).

   Minimum: _____   Q1: _____

   Median: _____   Q3: _____   Maximum: _____

3. Create a box plot for your group's data.

4. Calculate your group's IQR. _____

5. Compare your group's data with the data from another group.
   What can you say about the comparison based on the box plots and IQRs?

   _____

   _____

   _____

   _____

   _____

   _____

Math Boxes

1 Staci used $\frac{3}{5}$ of a $1\frac{1}{2}$-pound bag of flour to make pasta.

How many pounds of flour did she use?

Number sentence:

_____

SRB
32

2 Find each value for the prism represented by the net.

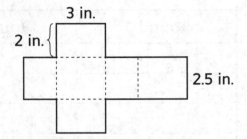

3 in.

2 in.

2.5 in.

a. Surface area: _____

b. Volume: _____

SRB
262-264

3 Square *ABCD* is formed from the following points:

A: $(0, -\frac{1}{2})$ B: $(0, 1\frac{1}{2})$

C: $(2, 1\frac{1}{2})$ D: $(2, -\frac{1}{2})$

What is the length of side *AB*?
Hint: Drawing a diagram may help.

Solution: _____

SRB
96

4 Record the solution set for each equation.

a. $x - 9 = 28$

_____

b. $x + 5 = x + 4$

_____

c. $x + 3 - 3 = x$

_____

SRB
208

5 **Writing/Reasoning** In Problem 2, explain how the net can be a useful tool for finding the surface area of the rectangular prism.

Math Boxes

1 Max borrowed $15 from his mom. Max's friend Alexis borrowed $17 from his dad.

Who has more debt?

Represent the situation using a number model and absolute value.

Number model: _____

Solution: _____

SRB
92-93

2 Use substitution to determine whether each equation is true or false.

a. $5x - 3 = 17$, if $x = 4$ _____

b. $6y^2 = 54$, if $y = 4$ _____

c. $t^2 - 14 = t + 7$, if $t = 5$ _____

SRB
207

3 Hakeem has $\frac{1}{4}$ pound of dried cranberries. Each morning he puts $\frac{1}{16}$ of a pound into his breakfast oatmeal.

How many breakfasts can he have with cranberries?

Solution: _____

SRB
32

4 Gabby plays a word game on her phone. Recently she scored 250 points, 250 points, 220 points, and 180 points. She wants her average to be above 250. Is she likely to have that average after her next game? Explain.

_____

_____

_____

SRB
285-288

5 Which expression matches the statement below?

Subtract five from the product of y and 6.

Fill in the circle next to the best answer.

○ **A.** $5 - 6y$

○ **B.** $6y - 5$

○ **C.** $5 * 6 * y - 5$

○ **D.** $6(y - 5)$

SRB
200-202

6 Describe the solution set for each inequality. Use set notation.

a. $7j > 77$

_____

_____

b. $17 \leq x + 9$

_____

_____

SRB
210-211

321

A Sweet Comparison
Using Unit Rates

Math Message

1 The sixth graders at Middlelands School are doing research on popular drinks. They took a survey and found that the three most popular drinks are the following:

- Thirsty Quench Sports Drink, which comes in 1-quart bottles

- Friendly Fruit Punch, which comes in 1-cup cartons

- Frosty Cola, which comes in 12-fluid-ounce cans

a. Order the drinks from least to greatest volume.

_____

b. Which drink do you think has the greatest concentration of sugar—that is, which has the most sugar per fluid ounce? Why?

_____

_____

Nutrition Labels for the Three Most Popular Drinks

Friendly Fruit Punch

Nutrition Facts

Serving Size 1 cup (8 fl oz)

Amount Per Serving	
Calories 112 Calories from Fat 0	
	% Daily Value*
Total Fat 0 g	0%
Saturated Fat 0 g	0%
Trans Fat 0 g	
Cholesterol 0 mg	0%
Sodium 2 mg	0%
Total Carbohydrate 32 g	8%
Dietary Fiber 0 g	0%
Sugars 32 g	
Protein 0 g	

Vitamin A 0%	Vitamin C 96%
Calcium 0%	Iron 0%

* Percent Daily Values are based on a 2,000 calorie diet. Your daily values may be higher or lower depending on your calorie needs.

Frosty Cola

Nutrition Facts

Serving Size 1 can (12 fl oz)
Serving Per Container 1

Amount Per Serving	
Calories 140	
	% Daily Value*
Total Fat 0 g	0%
Saturated Fat 0 g	0%
Trans Fat 0 g	
Cholesterol 0 mg	0%
Sodium 45 mg	2%
Total Carbohydrate 39 g	13%
Dietary Fiber 0 g	0%
Sugars 32 g	
Protein 0 g	0%

* Percent Daily Values are based on a 2,000 calorie diet. Your daily values may be higher or lower depending on your calorie needs.

Thirsty Quench Sports Drink

Nutrition Facts

32 fl oz (960 ml)
Servings per Container about 2.6

Calories 213

Amount/Bottle	%DV*	Amount/Bottle	%DV*
Total Fat 0 g	0%	Total Carb 59 g	7%
Sodium 400 mg	16%	Sugars 56 g	
Potassium 93 mg	2.5%	Protein 0 g	
Niacin	40%	Vitamin B6	40%
Vitamin B12	40%	Magnesium	†

† Not a significant source of calories from fat, saturated fat, trans fat, cholesterol, dietary fiber, vitamin A, vitamin C, calcium and iron.

* Percent Daily Values are based on a 2,000 calorie diet.

Use the nutrition labels on journal page 322 for the problems below.

2   Use the information from the nutrition labels to find the unit rates in the table.

	Unit Rates			
	Grams of Sugar (per cup)	Grams of Sugar (per fluid ounce)	Grams of Sugar (per 12 fluid ounces)	Grams of Sugar (per quart)
Fruit Punch				
Cola				
Sports Drink				

Use the unit rates to order the drinks from least to greatest concentration of sugar.

_____

3   Compare your order in Problem 2 to your prediction from the Math Message problem. Were there any surprises? Why or why not?

_____

_____

4   Ten teaspoons of sugar is 40 grams of sugar.

   a. Complete the ratio/rate table to find the number of teaspoons of sugar in 1 fluid ounce of each drink type. You do not have to use all the rows, or you may want to add rows.

   b. Find the number of teaspoons of sugar in 1 fluid ounce of each drink.

   Cola: _____ tsp sugar per fluid ounce

   Fruit punch: _____ tsp sugar per fluid ounce

   Sports drink: _____ tsp sugar per fluid ounce

Grams of Sugar	Teaspoons of Sugar
40	10

A Sweet Comparison
Using Unit Rates (continued)

5 The American Heart Association recommends that preteens and teenagers consume no more than 3–9 teaspoons of added sugar daily, depending on their age, size, and level of activity. If measured in whole-number fluid ounces, what is the largest serving size a preteen or teen could have of each drink within the 9-teaspoon limit?

Complete the ratio/rate table to calculate the sugar content for different serving sizes of Thirsty Quench. Use the space below to draw ratio/rate tables for Frosty Cola and Friendly Fruit Punch.

a. Thirsty Quench: _____

b. Frosty Cola: _____

c. Friendly Fruit Punch: _____

Thirsty Quench

Number of Fluid Ounces	Number of Teaspoons of Sugar
1	
2	
3	
4	
5	
10	
20	
21	

Try This

6 Jolene drinks 1 cup of Friendly Fruit Punch at breakfast, one 12-fluid-ounce can of Frosty Cola at lunch, and one 32-fluid-ounce bottle of Thirsty Quench Sports Drink during soccer practice.

By how much does she exceed the recommended daily limit of 9 teaspoons of sugar? _____

Math Boxes

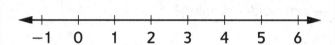

① Graph the solution set for $7x > 28$ on the number line.

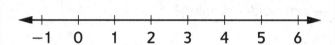

-1   0   1   2   3   4   5   6

SRB
210-211

② A school divides its rectangular garden among 25 classes. Each classroom gets a rectangular section measuring 4.5 feet by 2.5 feet.

If there is no extra space, what could the dimensions of the entire school garden be?

Solution: _____

SRB
32

③ Write at least two expressions that are equivalent to $4(x + 5)$.

SRB
206

④ Owen is 3 seconds behind the leader in a race, and Jordan is 5 seconds behind.

Who is farther behind the leader?

Solution: _____

Write a number model that represents the reasoning for your answer.

Number model: _____

SRB
92-93

⑤ Gracie hopes to win a prize for her watermelon at the state fair.

Last year's winning watermelon weighed $152\frac{1}{2}$ pounds.

Gracie's watermelon currently weighs $99\frac{3}{4}$ pounds.

How much heavier does her watermelon need be to tie last year's winner?

Solution: _____

SRB
32

⑥ Write at least three equivalent names for 9% in the name-collection box.

9%

SRB
56-58

Math Boxes

Comparisons with Unit Rates

Math Message

1 As of 2014, the women's world record for the 100-meter dash is 10.49 seconds. About how many meters per second was the record holder running?

_____

2 About how long would it take to run 400 meters if a runner maintained the same rate as the record holder in Problem 1? _____

3 The world-record time for the women's 400-meter dash is 47.60 seconds. How does this time compare to your answer in Problem 2?

_____

_____

4 Find the unit rate for each race in the table below.

Men's World Record Times in 2014		
Race Length	Time	Speed (meters per second, to the nearest hundredth)
100 m	9.58 sec	
200 m	19.19 sec	
400 m	43.18 sec	
800 m	1 min 40.91 sec	
1,000 m	2 min 11.96 sec	
1,500 m	3 min 26 sec	
2,000 m	4 min 44.79 sec	
3,000 m	7 min 20.67 sec	
5 km	12 min 37.35 sec	
10 km	26 min 17.53 sec	

5 What is the relationship between race length and the speed of the runner?

_____

Comparisons with Unit Rates (continued)

6 Predict what a graph for race lengths (horizontal axis) and speed (vertical axis) will look like.

_____

7 Plot the points for the graph.

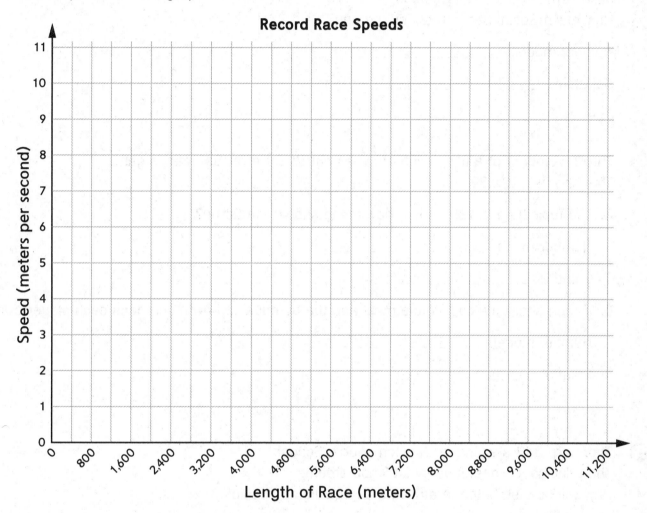

Record Race Speeds

Speed (meters per second)

Length of Race (meters)

8 How does the graph compare to what you predicted?

_____

9 What record speed (meters per second) would you expect for a 600-meter race?

_____

10 What record speed would you expect for an 8,000-meter race?

_____

327

Multiplying and Dividing

In the problems below, make sure your solution makes sense in each context. For example, money answers should be correct to the nearest cent.

1 The employee share of Social Security taxes is 6.2%.
Mike earned $3,276 as a busboy.
How much Social Security tax did he pay?

Number model: _____

Solution: _____

2 The 9 members of a baseball team went out to eat after Saturday's game.
The bill was $102.32.

   a. The typical tip is 18%. About how much should the tip be?

      Number model: _____

      Solution: _____

   b. If they share the cost of the meal and the tip equally, how much does each player owe?

      Number model: _____

      Solution: _____

3 There are 342 students at Newton Middle School.
The principal divides them into 15 homerooms.
How will he distribute the students as equally as possible?

Number model: _____

Solution: _____

4 There are 5 groups in the science lab.
Each group must receive the same amount of water for an experiment.
Ms. Hamilton has 3 containers with $2\frac{1}{2}$ liters of water in each.

   a. How much water does Ms. Hamilton have for the class? _____

   b. How much water should each group receive? _____

Learning about Conserving Water

Math Message

Olivia's science teacher, Ms. Palopoli, shared the following information:

- A leaky faucet drips 1 gallon of water in about 10,000 drips.

- The water flow of a standard showerhead is about 2.5 gallons per minute.

- The water flow of a standard bathroom sink faucet is about 2.2 gallons per minute.

- The water flow of a typical garden hose is about 11 gallons per minute.

1 Ms. Palopoli asked her students to calculate about how much water their families use. Olivia estimated her family's water use during the month of May as follows:

- Olivia's family typically took 4 showers per day. The showers had an average length of 8 minutes each. Olivia's family bathroom had a standard showerhead.

- Their bathroom sink leaked at the rate of 1 drip per second.

- The family typically watered the vegetable garden for about 5 minutes each day.

- The family typically brushed their teeth at the bathroom sink 8 times per day and left the water running for about 2 minutes each time.

Complete the table to show about how much water Olivia's family typically uses for the four listed categories per day and during a 30-day period. Include appropriate units for each table entry.

Source of Water Use	Daily Water Use	30-Day Water Use
Showering		
Brushing Teeth		
Leaking Faucet		
Watering the Garden		
Total		

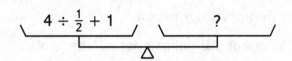

Math Boxes

① Write an equivalent numerical expression that will balance the pans. Use at least one operation in your expression.

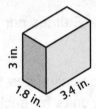

$4 \div \frac{1}{2} + 1$?

Expression: _____

SRB
217-218

② Compute.

a. $1.5(3.2 + 1.8) =$ _____

b. _____ $= 9(5.6 - 3.2)$

c. $0.06(9.5 + 1.75) =$ _____

SRB
204-205

③ Find the volume of this rectangular prism:

3 in.

1.8 in.   3.4 in.

Volume: _____

SRB
262

④ Mr. Dexter sells magazine subscriptions for $18 each. For each subscription he sells, he earns $8.

Last month he sold $900 worth of subscriptions.

How much did he earn? _____

SRB
43-44,
49-50

⑤ **Writing/Reasoning** Explain how you could use a ratio/rate table to answer Problem 4.

Representations for a Growing Pattern

Math Message

1   **a.** Use square pattern blocks to build the first three steps for the pattern below.

  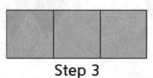

1 unit

Step 1   Step 2   Step 3

  b. In the space below, sketch Step 4 and Step 5 of the sequence.

Step 4   **Step 5**

2   **a.** Complete the table, and write an equation to represent the rule.

    Rule: _____

  b. Use substitution for one of the step values to show that your rule works for that step.

    _____

Step Number (x)	Perimeter (y) (units)
1	4
2	6
3	
4	
	12

---

3   Use the values in the table from Problem 2 as the x- and y-coordinates for points.

  a. List the points:

    _____

  b. Graph the points on the coordinate grid.

4   At Step 8, what is the perimeter? _____

5   Explain how you found your answer.

    _____

    _____

    _____

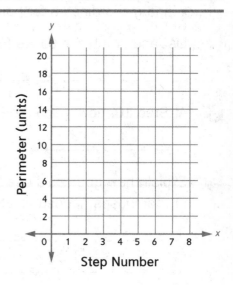

331

Representations for a Growing Pattern (continued)

6

Step 1   Step 2   Step 3

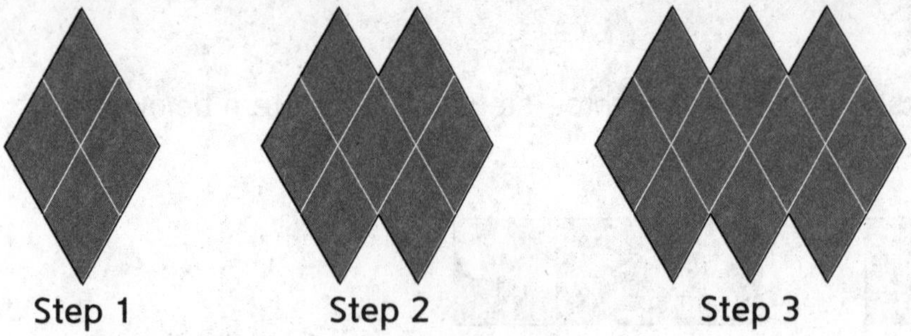

a. Describe what you would do to make the figure for Step 4.

_____

_____

b. Complete the table that represents the number of rhombuses for each step of the pattern.

c. Record an equation that represents the rule for the number of rhombuses in each step.

Rule: _____

d. Substitute one of the step values from the table to show that your rule works.

_____

Step Number (x)	Number of Rhombuses (y)
1	4
2	
3	
4	
5	
10	
100	
n	

7 Use the values in the table from Problem 6b as the x- and y-coordinates for points.

List and graph the points for Steps 1–5:

_____

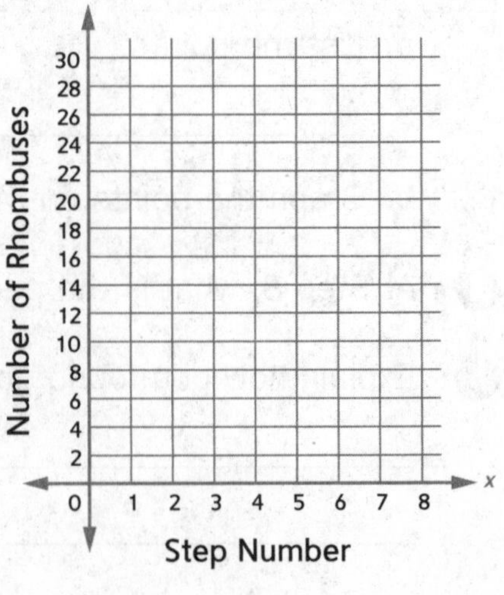

8 At Step 10, how many rhombuses do you need?

_____

9 Explain how you could use the graph to find the number of rhombuses needed at Step 10.

_____

_____

_____

Graph: y-axis labeled "Number of Rhombuses" with values 2 to 30; x-axis labeled "Step Number" with values 1 to 8.

Math Boxes

1 Nitrogen freezes, or becomes dry ice, at −78.5°C. Liquid nitrogen boils at −196°C.

Which temperature is colder?

_____

Write a number sentence to represent the temperature comparison.

_____

SRB
92-93

2 Tell whether each equation is true or false for the given value of the variable.

a. $10x^2 + 3 = 102$, if $x = 4$ _____

b. $288 = k * 2k$, if $k = 12$ _____

c. $7p ÷ 5 = 2p + 10$, if $p = 5$ _____

SRB
207

3 Robert bought a $1\frac{3}{4}$-pound bag of granola.
Each day, he eats $\frac{1}{8}$ pound for a snack.

For about how long will
his bag of granola last? _____

SRB
32

4 Lou ran 3 miles on Monday, 4 miles on Wednesday, and 2 miles on Friday.
He wants to run an average of 2 miles per day for the week.

How many miles does
he need to run
over the weekend? _____

SRB
205-208

5 Write an expression for each situation.

a. 8 less than the product of 10 and y

Expression: _____

b. 14 more than double z

Expression: _____

c. Half of the product of 4 less than t and 3

Expression: _____

SRB
200-202

6 Record the solution set for each problem as an inequality. Then record the solution set using set notation.

a. $x + 213 ≥ 319$ _____

b. $290 > h − 11$ _____

c. $1,004 ≤ 4 * m$ _____

SRB
210-211

Comparing Rates in an Ironman Triathlon

Math Message

An Ironman Triathlon consists of three stages, or legs.
Participants swim 2.4 miles, bike 112 miles, and run 26.2 miles.

1 The average participant takes 12 hours and 35 minutes to complete the race. If the average participant travels all three legs of the race at a constant speed, about how many minutes does it take to travel 1 mile? _____

2 The following statistics reflect the average times for completing each leg of the race.

Swim: 1 hour 16 minutes   Bike: 6 hours 25 minutes   Run: 4 hours 54 minutes

Approximately what is the average travel rate for each leg (in minutes per mile)?
Compute each answer to the nearest minute and record an equation in the table.

Leg of Race	Average Rate	Equation
Swimming		
Biking		
Running		

3 Complete the tables using the rates you found in Problem 2.

Swimming

Time (x) (minutes)	Distance (y) (miles)
	0.5
	1
	1.5
96	
	5
320	

Biking

Time (x) (minutes)	Distance (y) (miles)
	0.5
	1
	1.5
9	
15	
	10

Running

Time (x) (minutes)	Distance (y) (miles)
	0.5
11	
16.5	
	3
	5
110	

4 Explain how you know that the numbers you entered in the tables make sense.

_____

_____

_____

Comparing Rates in an Ironman Triathlon (continued)

5 Use the values in the tables from Problem 3 as the *x*- and *y*-coordinates for points.
Graph the points for each leg of the race on the coordinate grid.
Connect the points for each leg of the race using a different colored pencil.

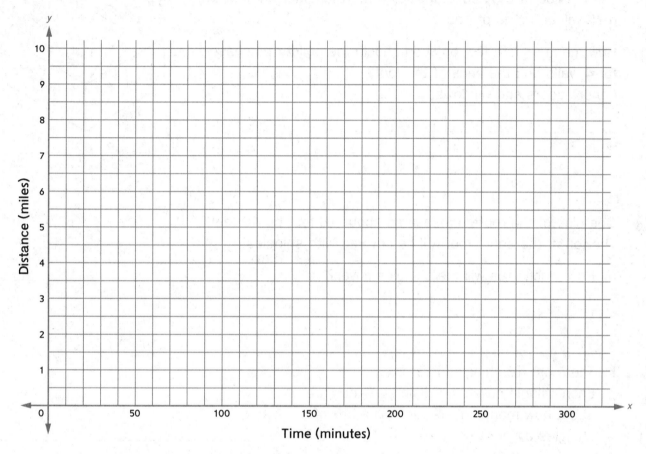

6 If you only had the graph, how could you tell which leg has the fastest rate?

_____

7 Why can the points on the graph for each rate be connected to make a straight line?

_____

_____

_____

8 Explain how you know which variable is independent and which is dependent.

_____

_____

335

Construction: Surface Area and Volume

A new house is being constructed with a rectangular prism for its base
and a triangular prism for its roof.

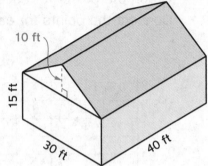

① The outside of the house, including the triangular ends of the
roof, will be made of wood.

How many square feet of wood should the builder purchase
for 4 walls and the ends of the roof?
(Ignore doors and windows.)

_____

② One side of the rooftop measures about 18 feet by 40 feet.
The top of the entire roof will be covered with shingles.

a. How much area will the shingles cover?   _____

b. There are 48 shingles in a box.
Each shingle covers a 9-inch by 9-inch area.
About how many square feet will a box of
shingles cover?   _____

c. How many boxes of shingles should the
builder purchase?   _____

③ To decide what size heater to install, the builder
needs to know the size of the interior of the
house. What is the volume of its interior
(including the space under the roof)?   _____

336

Math Boxes

Math Boxes

① Two stores sell digital cameras for about the same average price.

Digital Dan's prices have a mean absolute deviation of $25. Cate's Cameras have a mean absolute deviation of $60.

You need an inexpensive camera and your mom needs a professional camera for work. Where should you shop and why?

_____

_____

_____

SRB
294

② Kareem Abdul-Jabbar played a total of 57,446 minutes of game time in his NBA career. If he were to play those minutes consecutively, about how many 24-hour days is that?

Solution: _____

SRB
68

③ Use GCF and the Distributive Property to rewrite each addition problem as a multiplication expression.

a. $42 + 49$ _____

b. $70 + 126$ _____

c. $55 + 121$ _____

SRB
105,
204-205

④ Draw a net that represents a triangular prism.

SRB
263-264

⑤ **Writing/Reasoning** Explain how you calculated your answer to Problem 2. Include units in your explanation.

337

Math Boxes

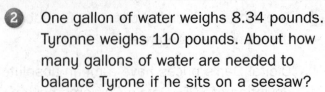

1 Graph the solution set.

$b - 3 \leq 2$

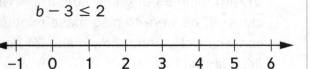

-1   0   1   2   3   4   5   6

SRB
210-211

2 One gallon of water weighs 8.34 pounds. Tyronne weighs 110 pounds. About how many gallons of water are needed to balance Tyrone if he sits on a seesaw?

Solution: _____

SRB
32

3 Which of the following expressions are equivalent to $3x + 12x$?

Circle ALL that apply.

A. $x(3 + 12)$

B. $15x$

C. $3(x + 4x)$

D. $15 + x$

SRB
206

4 Zach and Keisha are playing a trivia game in which correct answers earn one point and incorrect answers earn negative points. Zach's score is −20 and Keisha's score is −5. Who is winning?

Solution: _____

Write a number model that represents your reasoning.

_____

SRB
92-93

5 A snail slowly makes its way across a garden path. In the first hour, the snail goes $\frac{1}{2}$ of the distance. In the next 30 minutes, the snail goes $\frac{1}{2}$ of the remaining distance. In the next 15 minutes, the snail goes $\frac{1}{2}$ of the remaining distance. What fraction of the distance has the snail gone in 1 hour and 45 minutes?

Solution: _____

SRB
32

6 Write three equivalent names for 150%.

150%

SRB
56-58

Falling Objects

Math Message

The picture at the right was created based on a series of photographs taken at intervals of $\frac{1}{20}$ of a second.

Elapsed Time (sec)	Total Distance Fallen (ft)
$\frac{1}{20}$	0.04
$\frac{2}{20}$	0.16
$\frac{3}{20}$	0.36
$\frac{4}{20}$	0.64
$\frac{5}{20}$	1.00
$\frac{6}{20}$	1.44
$\frac{7}{20}$	1.96
$\frac{8}{20}$	2.56
$\frac{9}{20}$	3.24
$\frac{10}{20}$	4.00

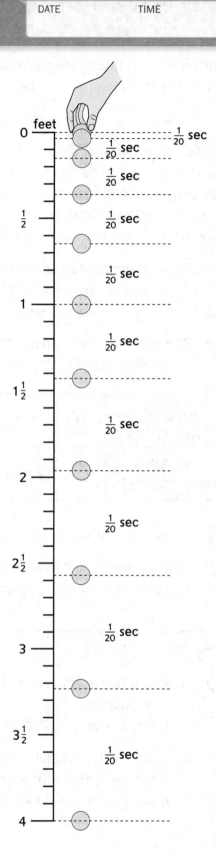

1. How far did the ball fall during the first $\frac{1}{4}$ second? _____

2. How far had it fallen after $\frac{1}{2}$ second? _____

3. a. Check the statement that you believe is true.

   ☐ A ball falls at a constant (even) speed.

   ☐ As a ball falls, it picks up speed.

   b. Explain your choice in Problem 3a.

   _____

   _____

   _____

339

What Happens When an Object Is Dropped?

Galileo was an Italian physicist who lived from 1564 to 1642. His work led to the following rule for the distance traveled by a freely falling object:

    Distance traveled in feet = 16 * square of the elapsed time in seconds

Written as a formula, the rule is $d = 16 * t * t$, or $d = 16 * t^2$, where d is the distance traveled by the object in feet and t is the time in seconds that has elapsed since the object started falling. For example, after 1 second an object will have traveled 16 feet (16 * 1 * 1); after 2 seconds, it will have traveled 64 feet (16 * 2 * 2).

Galileo's formula really applies only in a vacuum, where there is no air resistance to slow an object's fall. However, it is a good approximation for the fall of a dense object, such as a bowling ball, over a fairly short distance.

① The following table shows the approximate distance traveled by a freely falling object at 1-second intervals. Use Galileo's formula to complete the table.

Elapsed Time (sec) t	0	1	2	3	4	5	6	7	8	9
Distance (ft) $16 * t^2$	0	16	64	144						

② Graph the data from the table above onto the grid on journal page 341.

③ Use the graph to estimate the number of seconds it takes an object to fall 500 feet, ignoring air resistance.   About _____ seconds

---

Try This

④ **a.** About how many seconds would it take an object to fall 1 mile, ignoring air resistance? Use your calculator.
(*Hint*: 1 mile = 5,280 feet)   About _____ seconds

b. In the table, evaluate Galileo's distance formula for at least two other values of t (time).

Elapsed Time (sec) t		
Distance (ft) $16 * t^2$		

What Happens When an Object Is Dropped? (continued)

Lesson 7-10

DATE

TIME

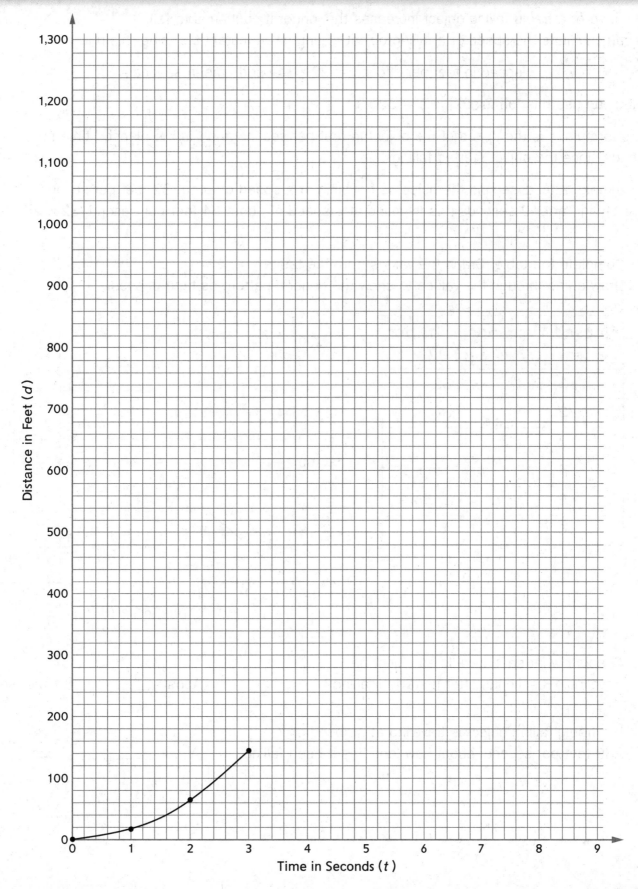

What Happens When an Object Is Dropped? (continued)

The speed of a freely falling object increases the longer it continues to fall.
You can calculate the speed of a falling object at any given instant by using this rule:

speed in feet per second = 32 * elapsed time in seconds

Or you can use this formula: $s = 32 * t$

In the formula, s is the speed of the object in feet per second and t is the elapsed time in seconds since the object started falling.

For example, after 1 second the object is traveling at a speed of about 32 feet per second (32 * 1), and after 2 seconds it is traveling at a speed of about 64 feet per second (32 * 2).

5 Complete the table. Graph the values from the table.
Draw a line through the values to show the general pattern as time passes.

Elapsed Time (sec) t	Speed (ft per sec) 32 * t
1	32
2	64
3	
4	
5	
6	

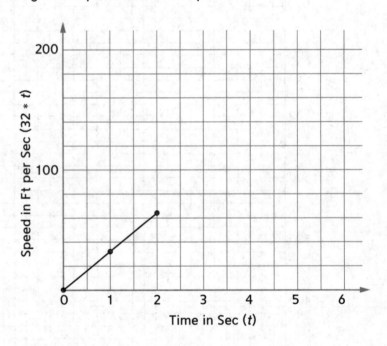

6 At about what speed would an object be
traveling 18 seconds after it started falling?   About _____ ft per sec

7 If a freely falling object is traveling at a speed of
290 feet per second, about how long has it been falling?   About _____ sec

342

Introducing Mystery Graphs

Math Message

The graphs below show how long it takes Dillon to get to school three different ways.

A

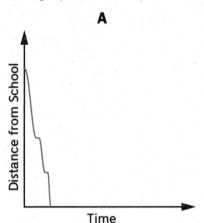

B

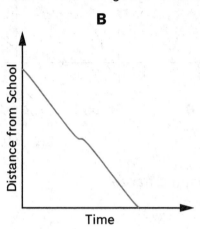

C

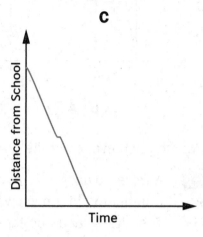

1. Match each description to the graph above that could show Dillon's distance from school over time.
   Record the letter of the graph next to the statement.

   a. Dillon walks to school slowly.   _____

   b. Dillon's brother drives him to school.   _____

   c. Dillon rides his bike to school.   _____

2. Explain how you made your choice for Problem 1a. _____

   _____

Complete the graph to represent the situation.

3. Monika filled a cup with cocoa. She drank half of it. She refilled the cup. Then she drank all of the cocoa.

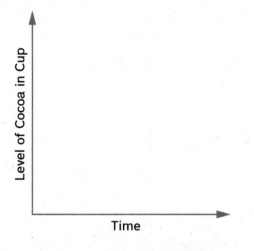

4. Shay walked up one side of a hill at a steady pace, ran at a constant speed down the other side, and then returned to walking at a slightly faster steady pace.

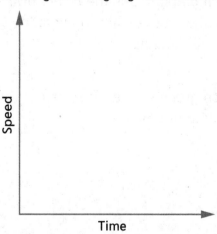

343

More Mystery Graphs

Each of the events described below is represented by one of the following graphs.

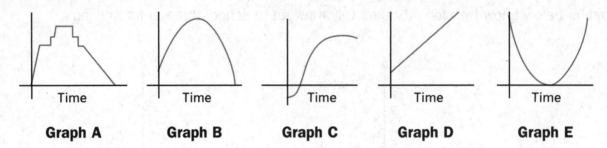

Match each event with its graph.

1 A frozen dinner is removed from the freezer.
It is heated in a microwave oven.
Then it is placed on the table.

Which graph shows the temperature
of the dinner at different times?   Graph _____

2 Satya runs water into his bathtub.
He steps into the tub, sits down, and bathes.
He gets out of the tub and drains the water.

Which graph shows the height of water
in the tub at different times?   Graph _____

3 A baseball is thrown straight up into the air.

a. Which graph shows the *height* of the ball
from the time it is thrown until the time it
hits the ground?   Graph _____

b. Which graph shows the *speed* of the ball
at different times?   Graph _____

Using Ratios

Solve the problems. Include appropriate units in your answers.
Hint: Tape diagrams or ratio/rate tables may be helpful.

1 Abner types an average of about 8 words in 6 seconds.
At that rate, how many words can he type in 5 minutes?   _____

2 Jorge earns $45 for 5 hours of work. He is saving up
to buy a bike that costs $350.
At that rate of pay, how many hours will he have to
work in order to earn enough money to buy the bike?   _____

3 Scholastic reports that the U.S. edition of the *Harry Potter*
series has 4,224 pages.
Delna read the first 40 pages in 75 minutes. At that rate,
how many hours will it take her to read all seven books?   _____

Try This

4 The first group to arrive at the sixth-grade square-dancing class
was made up of 4 girls and 1 boy. As each new group arrives,
Bill notices that they always arrive in a ratio of 2 girls to 1 boy.

a. As more groups with the ratio of 2 girls to 1 boy arrive,
is the ratio of girls to boys increasing or decreasing?   _____

b. If the ratio of newly arriving groups is always 2 girls to
1 boy, does the ratio of girls to boys ever reach 2 to 1?
Explain your reasoning.

_____

_____

_____

_____

_____

Math Boxes

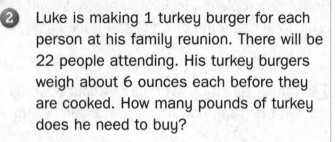

1 Fisher School students are wearing pedometers for one day to track their steps. The mean average deviation for fourth-grade students is 85 steps. The mean average deviation for fifth-grade students is 120 steps. What can you say about the walking habits of students in the two grades?

_____

_____

_____

SRB
294

2 Luke is making 1 turkey burger for each person at his family reunion. There will be 22 people attending. His turkey burgers weigh about 6 ounces each before they are cooked. How many pounds of turkey does he need to buy?

Number model: _____

Solution: _____

SRB
68

3 Use GCF and the Distributive Property to rewrite each subtraction problem as a product.

a. 88 − 77 = _____

b. 75 − 55 = _____

c. 68 − 12 = _____

SRB
105,
204-205

4 Sketch a net that represents a cube.

SRB
263-264

5 **Writing/Reasoning** Explain how you know that your net in Problem 4 could make a cube.

Math Boxes

1   Write an equivalent numerical expression that will make the equation true. Use at least one operation in your expression.

_____ $= 5^2 - 3$

SRB
206

2   Compute.

a. $12.7(0.2 + 0.8) =$ _____

b. _____ $= 5.02(5 - 3.2)$

c. $0.001 * (0.5 + 1.75) =$ _____

SRB
204-205

3   Find the height (h) of the aquarium shown if the volume of the aquarium is 6,912 in.³.

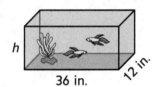

36 in.   12 in.

$h =$ _____

SRB
262

4   At Kozminski School, the ratio of weeks of school to weeks of vacation is 9 to 4. How many weeks of vacation do students get in 1 year?

Solution: _____

SRB
43-48

5   **Writing/Reasoning** Describe how knowing the formula for volume might help you find the height of the aquarium in Problem 3.

Analyzing Garden Ratios

Math Message

1 Francisco is considering options for a vegetable garden. His garden club suggests that lettuce seeds should be planted 6 inches apart.

   a. Draw a diagram to show the maximum number of lettuce plants that will fit in a garden row that is 5 feet long.

   b. What is the maximum number of lettuce plants that fit in one row? _____

2 Traditionally, gardeners plant in rows.
They leave 3 feet between the rows to allow room to tend the plants.

   a. Francisco has one area that is about 5' by 7'. Draw and label a diagram to show the maximum number of lettuce plants he can grow in this area.

   b. How many lettuce plants could he plant in this area? _____

   c. What is the ratio of plants to square feet for this garden? _____

There is another gardening method called "square-foot gardening."

Square-foot garden beds measure no more than 4 feet wide and are subdivided into a grid of 1 ft by 1 ft squares.

The diagram shows how many lettuce plants would fit in a 3 ft by 1 ft space using the square-foot gardening method.

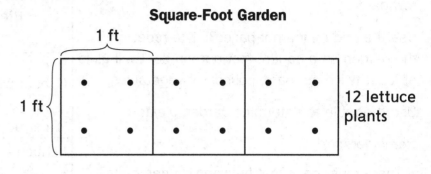

Square-Foot Garden

1 ft

1 ft

12 lettuce plants

The lettuce plants are at least 6 inches from each other and 3 inches from the border. Each square could be planted the same way.

3   **a.** To grow at least as many lettuce plants as you found for Problem 2b, how many 1' by 1' squares does Francisco need?_____

     b. How many plants will be in the square-foot garden in Part a? _____

     c. What is the ratio of plants to square feet for this garden? _____

4   What is the unit rate for how many plants can be planted per square foot in each type of garden? *Hint:* Use your ratios in Problems 2c and 3c to calculate the unit rates.

Row Garden	**Square-Foot Garden**
(number of plants per square foot)	(number of plants per square foot)
_____	_____

5   One lettuce plant weighs about 2 pounds. For each type of garden, estimate the yield (pounds) per square foot for lettuce plants.

Row Garden	**Square-Foot Garden**
(number of pounds per square foot)	(number of pounds per square foot)
_____	_____

6   Compare the two gardening methods.
Circle the statement below that best describes the relationship between the garden yields.

     A.   A row garden will yield about 2 times as much as a square-foot garden.

     B.   A square-foot garden will yield about 2 times as much as a row garden.

     C.   A row garden will yield about 4 times as much as a square-foot garden.

     D.   A square-foot garden will yield about 4 times as much as a row garden.

Doubling Garden Yield per Square Foot

The diagram below shows the layout of a row garden.

Use the grid on journal page 351 to redesign the garden as a square-foot garden so it will yield at least twice as many pounds of vegetables.

Use the table to determine garden yields.

Reminders:

- There must be 3 feet between garden beds.

- Garden beds can be no more than 4 feet wide.

- Garden beds must be 3 feet from the edges of the garden.

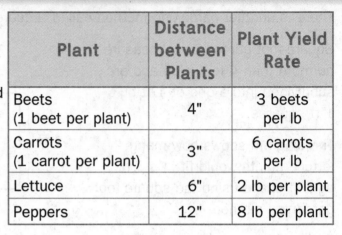

Plant	Distance between Plants	Plant Yield Rate
Beets (1 beet per plant)	4"	3 beets per lb
Carrots (1 carrot per plant)	3"	6 carrots per lb
Lettuce	6"	2 lb per plant
Peppers	12"	8 lb per plant

① Based on the table, label each row in the diagram with an estimate of its yield.

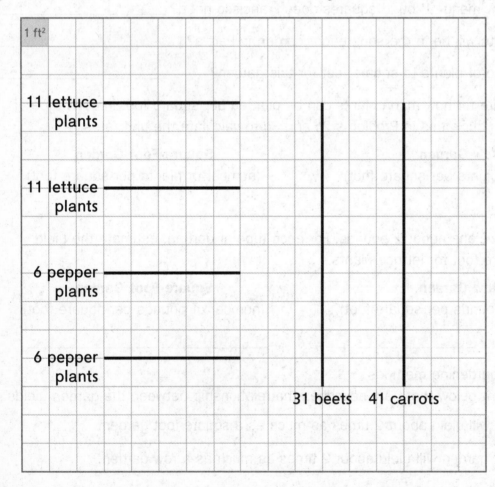

11 lettuce plants

11 lettuce plants

6 pepper plants

6 pepper plants

31 beets   41 carrots

② Estimate the total yield for the row garden (in pounds). _____

③ Record the rate of yield in pounds per square foot. _____

Doubling Garden Yield per Square Foot (continued)

④ Label your diagram below to show what kind of plants are in each garden bed. Use the yield rates listed on journal page 350 to label each garden bed with its yield (in pounds).

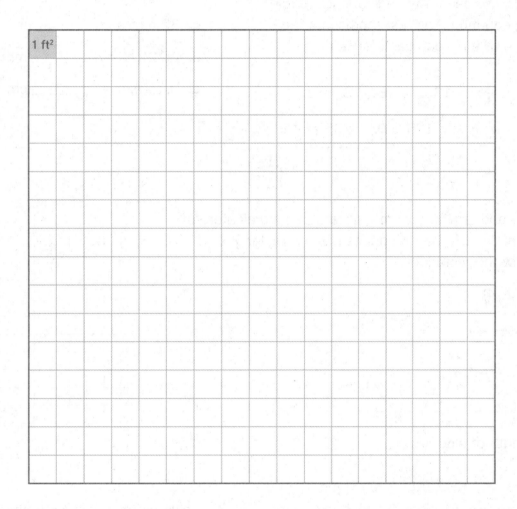

1 ft²

⑤ Estimate the total yield for the square-foot garden (in pounds).

_____

⑥ Record the rate of yield in pounds per square foot.

_____

Math Boxes

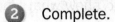

① Shamar wants to buy a video game system. He has \$135, which is $\frac{1}{3}$ of the price of the system.

Let x represent the original price. Circle the equation that best represents the price of the video game system.

A. $\frac{1}{3}x = 135$ **B.** $3x = 135$

C. $x + \frac{1}{3} = 135$ **D.** $135 = \frac{1}{3} \div x$

SRB
200-202

② Complete.

a. _____ m = 368 mm

b. _____ cm = 0.245 m

c. 32 mm = _____ m

d. 45.2 cm = _____ mm

e. 0.25 mm = _____ cm

SRB
68-70

③ Plot and label the points. Locate the coordinates for point D such that the points make a rectangle. Draw the rectangle.

A: (2, 4)

B: $(2, -\frac{1}{2})$

C: $(-3\frac{1}{2}, -\frac{1}{2})$

D: (_____, _____)

Length of side AB: _____

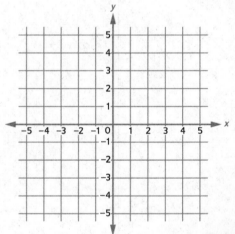

SRB
95-96

④ Name two solutions for each inequality.

a. $1.1 - 0.38 < w$ _____

b. $10\frac{1}{3} - 6\frac{2}{3} \geq g$ _____

SRB
210-211

⑤ Solve the pan-balance problem.

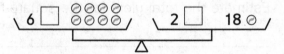

One ☐ weighs as much as

_____ marbles.

SRB
217-218

352

1 Complete the table. Write a rule.

Rule: _____

x	y
2	0.7
3	1.05
4	
	3.5

SRB
221

2 Evaluate.

a. $15 - 3.3 * 4 =$ _____

b. $0.01 + 0.01 * 10 + 0.01 =$ _____

c. _____ $= 0.5(3.2 - 1.7)$

d. $0.8 + 0.9 \div 0.3 =$ _____

SRB
128, 134,
154

3 The spreadsheet shows how Jonas spent his money for the first quarter of the year.

a. In which cell is the greatest amount that Jonas spent?

_____

b. Calculate the values for the cells E2, E3, and E4 and enter them in the spreadsheet.

	A	B	C	D	E
1	Month	Food	Movies	Music	Total
2	January	$38.50	$34.00	$62.50	
3	February	$29.45	$28.70	$26.89	
4	March	$34.90	$41.86	$48.30	

c. Circle the correct formula for calculating the amount of money Jonas spent in February.

D1 + D2 + D3   D3 − C2 + C3   B3 + C3 + D3

SRB
229-230

4 **Writing/Reasoning** Explain how you determined the rule for the table in Problem 1.

Scaling Down a Gallery Wall

Math Message

1 The scale drawing on journal page 355 represents an art gallery wall. The wall in the gallery is 9 feet tall and 8 feet wide. Each grid square is $\frac{1}{4}$ inch by $\frac{1}{4}$ inch (with an area of $\frac{1}{16}$ in.²).

What scale was used to make the drawing? _____

2 Justify your answer.

_____

_____

_____

3 Use the scale you found in Problem 1 and the scale drawing on journal page 355 to complete the table.

Letter	Scaled Dimensions (height × width)	Actual Dimensions (height × width)
O	$\frac{3}{8}$" × $\frac{9}{16}$"	6" × 9"
A		
B		
C		
D		
E		
F		

4 Add rectangles for two more pieces of art to the scale drawing of the gallery wall. Use the dimensions listed in the table below and complete the table.

Letter	Scaled Dimensions (height × width)	Actual Dimensions (height × width)
G		18" × 9"
H		12" × 18"

354

Our Art Gallery

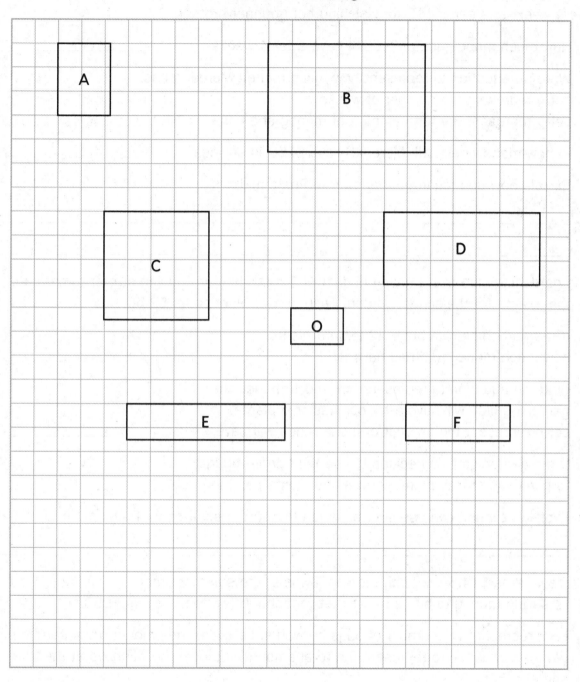

Spending and Saving

1 Meredith has $100 in her savings account.
Every week, she earns $12 for watching her younger brother.

  a. How much money does Meredith have after 4 weeks? _____

  b. Write an equation to represent how much money Meredith has.
  Let *b* represent the number of weeks.
  Let *y* represent the total amount of money she has. _____

  c. In how many weeks will Meredith have $250 in savings? _____

  d. Explain how you found your answer to Problem 1c.

  _____

  _____

2 The gardening club has a yearly budget of $200. They are buying starter tomato and
pepper plants for the school garden. Each starter plant costs $2.50.

  a. If the gardening club buys 22 tomato plants,
  how much money will they have left? _____

  b. Write an equation to represent how much money the
  gardening club will have after purchasing *z* plants.
  Let *h* represent the amount of money they will have. _____

  c. If the gardening club decides to use their entire budget
  to purchase starter plants, how many plants can they buy? _____

  d. Explain how you can use your equation to find the solution for Problem 2c.

  _____

3 Michael earns $14.50 per day mowing lawns in the summer time.
He is saving money to buy a new tablet for school. The tablet costs $250.

  a. Let *t* represent the number of days he works. Let *w* be the amount of money he has.
  Write an equation to represent the total amount of money he will have at the
  end of *t* days.

  _____

  b. Use your equation to figure out how many days
  he will have to work to have enough money for the tablet. _____

Math Boxes

Math Boxes

① Tyrell bought 2 pairs of jeans for $34.99 each and 2 T-shirts for $19.99 each. Let b represent the difference in price. Which number model(s) can be used to calculate how much more he spent on jeans than on T-shirts? Circle all that apply.

A. $2(34.99 + 19.99) = b$

B. $b = 2 * 34.99 - 19.99$

C. $2(34.99 - 19.99) = b$

D. $b = 2 * 34.99 - 2 * 19.99$

SRB
200-202

② Complete.

a. _____ m = 5,670 mm

b. _____ cm = 2.24 m

c. 362 mm = _____ m

d. 952 cm = _____ mm

e. 25 mm = _____ cm

SRB
68-70

③ Plot and connect points to make polygon *ABCD*.

A: (−1, 3)

B: (3, 3)

C: (2, 1)

D: (−2, 1)

Length of side *CD*: _____

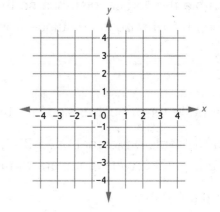

SRB
95-96

④ Name two solutions for each inequality.

a. $n \geq 3\frac{11}{16} + 4\frac{1}{2}$ _____

b. $k < 6\frac{2}{5} * 15$ _____

SRB
210-211

⑤ Solve the pan-balance problem.

3x + 3   15 + x

$x =$ _____

SRB
217-218

Math Boxes

Stretching Figures on a Coordinate Grid

Math Message

Use the Original Fish artwork from *Math Masters,* page 332 and the table of ordered pairs on journal page 359 to answer the questions.

1. Find the lengths for triangle *WVX*. Use $\overline{WV}$ as the height.

   a. The height of triangle *WVX* has what length?

   Number model: _____   Answer: _____

   b. The base of triangle *WVX* has what length?

   Number model: _____   Answer: _____

2. **a.** Points *W* and *X* are the same vertical distance from point *V*.
   What is the vertical distance from point *V* for these points? _____

   b. Double the vertical distance so that points *W* and *X* are twice as far below *V*.
   Keep their horizontal distance from point *V* the same. List the new ordered pairs.

   W: _____   *X:* _____

3. Write the letter of your assigned artwork. Artwork: _____

   Using the table on journal page 354, describe how the dimensions of
   your assigned artwork compare to the dimensions of Artwork O, the original artwork.

   Use ratio notation.

   Ratio of heights for Artwork _____ to Artwork O: _____

   Ratio of widths for Artwork _____ to Artwork O: _____

4. Describe how the dimensions of your assigned artwork will differ from Artwork O.

   _____

   _____

   _____

   _____

Stretching Figures
on a Coordinate Grid (continued)

5 Complete the table by recording the new ordered pairs for your assigned artwork.

Points in Original Fish	Original Ordered Pairs for Original Fish	New Ordered Pairs for Assigned Artwork _____
M	(−10.5, −2)	
N	(−2, 6)	
P	(4, −1)	
Q	(10, 3)	
R	(7.5, −2)	
S	(10, −7)	
T	(4, −3)	
U	(−2, −10)	
V	(−2, −2)	
W	(−2, −5.5)	
X	(1, −5.5)	
Y	(−6, −1)	

6 Use the ordered pairs to draw your artwork on the coordinate grid you assembled from *Math Masters*, pages 333–336.

7 Explain how you can figure out where on the wall to hang your assigned artwork.

_____

_____

_____

_____

_____

Celestial Body Data and Scale

Math Message

1 The table provides estimates of the **diameters** (distance from one
side to the other through the center) of various celestial bodies.

The estimates are in kilometers.

In the last column, compare each diameter with the diameter of Earth.

Celestial Body	Average Diameter (km)	Estimated Diameter Compared to Earth's Diameter
Mercury	4,900	About $\frac{3}{8}$ or $\frac{1}{2}$
Venus	12,000	About the same
Earth	13,000	
Mars	6,800	
Jupiter	140,000	
Saturn	120,000	
Uranus	51,000	
Neptune	49,000	
Pluto	2,300	
Sun	1,400,000	

2 The scale factor we will use for the diameter is 1 cm = _____.

3 My team's planet is _____.

Its average diameter is about _____.

Celestial Body Data and Scale
(continued)

Use the scale from Problem 2 on journal page 360.

4   In the scale model, the diameter of my planet will be _____.

5   **a.** The diameter of the Sun in the model will be _____.

    b. The diameter of the Sun in meters in the model is _____.

6   Share information across teams to complete the table.

7   What ratio comparison would you use to describe the relationship between the largest celestial body and Earth?

_____

_____

_____

Celestial Body	Diameters of Models Scale: 1 cm = _____
Mercury	
Venus	
Earth	
Mars	
Jupiter	
Saturn	
Uranus	
Neptune	
Pluto	
Sun	

8   What are some other ratio comparisons that might be useful in making your model?

_____

_____

_____

_____

_____

_____

Making a Scale Model of Your Planet

To make a 2-dimensional scale model of your planet, your team needs a pencil, a ruler, scissors, tape, a compass, string, and colored construction paper.

If possible, use the chart at the right to select the color(s) of paper for your planet.

Planet	Color
Mercury	Orange
Venus	Yellow
Earth	Blue, brown, green
Mars	Red
Jupiter	Yellow, red, brown, white
Saturn	Yellow
Uranus	Green
Neptune	Blue
Pluto	Yellow

Step 1 Use a ruler to draw a line segment equal in length to the diameter your planet should have in the model. If you are modeling Jupiter or Saturn, you may need to tape 2 sheets of paper together.

Step 2 Find the **midpoint** (middle) of this line segment and mark a dot there.

Step 3 Use a compass to draw a circle. The center of the circle should be at the midpoint you marked. Put the point of the compass on the dot. Put the pencil on one endpoint of the line segment and draw the circle.

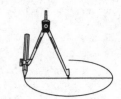

If your compass is too small, tie a string around a pencil near the point. Hold the point of the pencil on one endpoint of the line segment. Pull the string tightly and hold it down at the dot (midpoint) on the line segment. Keeping the string tight, swing the pencil around to draw a circle.

Step 4 Cut out and label the circle.

Step 5 Share your work with other teams.

Solving Equations Different Ways

Use a bar-model strategy at least once and an inverse-operation strategy at least once.

1 Amir spent $17.50 at the comic book store. He has $21.93 left.
How much money did he have before he bought the comics?

Define your variable. _____

Equation: _____

Show the steps for solving your equation. _____

Solution: _____

Check: _____

2 Liam sent 80% as many texts as his sister Diane.
If Liam sent 260 texts this month, how many did his sister send?

Define your variable. _____

Equation: _____

Show the steps for solving your equation. _____

Solution: _____

Check: _____

3 $4m + 4 = 2m + 6$

4 $u + 87 = 9u + 7$

$m =$ _____

$u =$ _____

Math Boxes

1 Write a rule. Fill in the missing numbers.

Rule: _____

x	y
2	$\frac{1}{2}$
3	$1\frac{1}{2}$
	$4\frac{1}{2}$
10	

SRB
221

2 Evaluate.

a. $8 + (4.5 - 0.9) - 2 =$ _____

b. $0.01 * 100 + 0.9 =$ _____

c. _____ $= 2.5(1.2 + 1.8)$

d. $1.5 \div 0.3 + 0.7 * 10 =$ _____

SRB
128, 134,
154

3 Darin charges $5 per hour to babysit on weekdays and $7 per hour on weekends. The spreadsheet is a record of the babysitting Darin did during one week.

	A	B	C
1	Day of the Week	Number of Hours	Earnings
2	Monday	4	
3	Wednesday	2	
4	Saturday	5	
5	Total		

a. Complete the spreadsheet.

b. Which cell contains the number of hours Darin worked on Saturday? _____

c. Circle any formulas Darin could use to calculate his total earnings.

C2 + C3 + C4   B2 + B3 + B4   (5 * B2) + (5 * B3) + (7 * B4)

SRB
229-230

4 **Writing/Reasoning** Explain how you determined a formula that would work for Problem 3.

364

Area of Our Classroom

Math Message

1 Record the classroom dimensions measured by your classmates.
Make a scale drawing of your classroom. Label your dimensions.

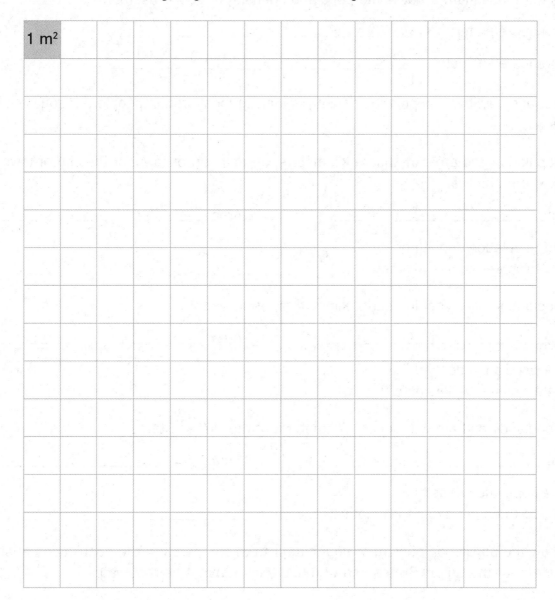

1 m²

2 Find the area of your classroom floor in square meters. _____

3 How many people, including students and teacher(s), are in your classroom? _____

4 If everyone in your class had an equal amount of floor space, about how
many square meters would each person get?
Find the answer to the nearest thousandth of a square meter. _____

365

Calculating Population Density

1 Record a number model for calculating the population density in your classroom.

_____

2 To the nearest thousandth, what is the population density of your classroom?

People per square meter: _____

People per square kilometer: _____

For cities, states, and countries, population density is usually measured in people per square mile or square kilometer.

3 Find the population and area, in square kilometers, of your city or town (or the city or town nearest to where you live).

Population: _____   Area: _____

What is the population density of
your city or town per square kilometer?   _____

4 Find the population and area, in square kilometers, of your state.

Population: _____   Area: _____

What is the population density
of your state per square kilometer?   _____

5 Find the population and area, in square kilometers, of the United States.

Population: _____   Area: _____

What is the population density
of the United States per square kilometer?   _____

6 Describe how the population density of your city or town compares to the population density of your state or of the United States and why you think it might be that way.

_____

_____

_____

Calculating Population Density (continued)

7 Compare the population density of your classroom to your city/town, state, and the United States.

    a. Which has the highest concentration of people? _____

    b. What might account for the high density? _____

    _____

    _____

8 What areas might have low population densities? _____

_____

Justify your answer. _____

_____

_____

9 Monaco is the country with the highest population density, with an area of 2 km^2 and a population of 30,500 people. Mongolia has the lowest population density, with an area of 1,553,556 km^2 and a population of 3,226,516 people.
Find the population densities per square kilometer.

Monaco: _____   Mongolia: _____

What do you know about Monaco and Mongolia that might help explain this difference?

_____

_____

_____

_____

Try This

10 A football field is 110 m long (from goal post to goal post) and 48.75 m wide.
How many people would have to be on a football field to equal the population densities of the locations listed below? Round your answer to the nearest hundredth.

Your classroom: _____   Monaco: _____

United States: _____   Mongolia: _____

Math Boxes

1 The Park and Shop parking lot has 14 rows of parking spots. Each row has the same number of parking spots. There are 154 total parking spots in the lot. How many cars can park in each row?

Number model: _____

Solution: _____

SRB
32

2 Use the general pattern to compute the following products mentally.

$$x(y - z) = (x * y) - (x * z)$$

Example: $8 * 99 = 8(100 - 1)$
$$= 8(100) - 8(1) = 792$$

a. $5 * 97$ _____

_____

b. $12 * 18$ _____

_____

SRB
204-205

3 Solve using a bar model.

$$4x + 10 = 16 + 2x$$

Solution: _____

SRB
216

4 Graph the solution set for $x > -2$ on the number line below.

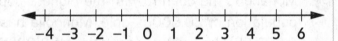

−4 −3 −2 −1 0 1 2 3 4 5 6

SRB
210-211

5 Malynn made a cubic box out of cardboard. What is the area of all the cardboard she used?

$10\frac{1}{2}$ in.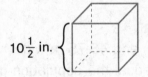

Area: _____

SRB
263-264

6 Use substitution to determine which numbers are solutions to the inequality.

Circle all that apply.

$$2y - 7 \leq 10$$

A. 8   **B.** 9

C. 7.5   **D.** 4

SRB
210-211

368

Making a Mobile

Math Message

Number of Books Read

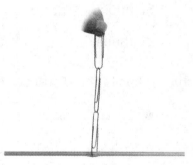

1 **a.** Draw the balance point on the dot plot to indicate where the mean is.

b. Add three more dots while keeping the two sides balanced.

Assemble the equipment your group needs to make mobiles:

- three straws • scissors • tape • ruler • a single-hole punch

- paper clips (small and large sizes) • heavy construction paper or cardstock

2 **Making a Top Layer**

- Mark the exact middle point on one of your straws.

- Make a chain of three large paper clips.

- Slide one end of your paper-clip chain onto the straw.

- Tape one end of the paper-clip chain to your center mark. This is the top layer of your mobile.

3 **Making Mobile 1**

- Make two rectangles with heavy paper. The ratio of the area of one rectangle to the other should be 1 : 2. Their weights will have approximately the same ratio as their areas.

- Punch a hole in each rectangle. Attach a small paper-clip chain to each rectangle. Use the same number of paper clips in each chain.

- Slide the ends of the paper-clip chains over opposite ends of the top layer. Move the chains around until the mobile is in balance.

- Tape the paper-clip chains to the straws in the locations that balance the mobile.

- Record your measurements on the diagram.

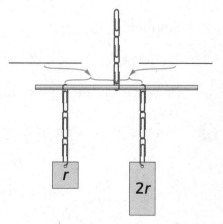

Try This

4 Design your own mobile so that it will balance. Make multiple layers by adding straws (and more paper clips) to the paper-clip chains instead of the rectangles.

Solving Mobile Problems

The mobile shown in each problem is in balance.
The **fulcrum** of the mobile at the right is the center point of the rod.
A mobile will balance if $W * D = w * d$.

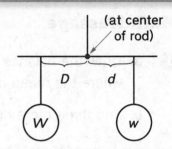

(at center of rod)

Write and solve an equation to answer each question.

1 What is the distance from the fulcrum to the object on the right of the fulcrum?

W = _____   D = _____

w = _____   d = _____

Equation: _____

Solution: _____

Distance: _____ units

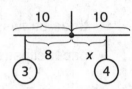

2 What is the distance from the fulcrum to each of the objects?

W = _____   D = _____

w = _____   d = _____

Equation: _____

Solution: _____

Distance on the left of the fulcrum: _____ units

Distance on the right of the fulcrum: _____ units

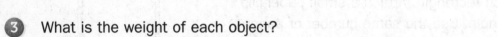

3 What is the weight of each object?

W = _____   D = _____

w = _____   d = _____

Equation: _____

Solution: _____

Weight of the object on the left of the fulcrum: _____ units

Weight of the object on the right of the fulcrum: _____ units

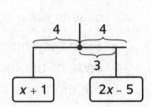

Graphing and Comparing Rates

Find the distances traveled by two trains.

1   Complete the tables.

Train A	
Time (hr)	Distance (mi)
0	
1	30
2	
3	
4	

Train B	
Time (hr)	Distance (mi)
0	
1	45
2	
3	
4	

2   What is the average speed for each train—that is, how many miles per hour is each train traveling on average?

Train A: _____   Train B: _____

3   Use the information in the tables to make a graph of the distance each train travels over time. Connect the points and label the lines to show which train each represents.

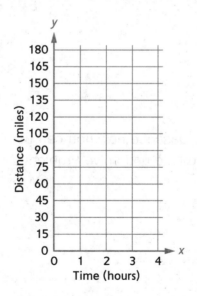

Try This

4   Describe a relationship between the speeds of the two trains.
    How fast does Train B travel compared to Train A? (*Hint:* Ratios might be helpful.)

_____

Math Boxes

1 Calculate the volume of the rectangular prism shown below.

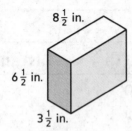

$8\frac{1}{2}$ in.

$6\frac{1}{2}$ in.

$3\frac{1}{2}$ in.

Volume: _____

SRB
262

2 One penny weighs about $\frac{1}{10}$ of an ounce. The school penny drive collected 1,275 ounces of pennies.

How much money did they raise?

Solution: _____

SRB
68-70

3 Chen ran 2 miles more than $\frac{2}{3}$ of the distance that Kayla ran. If Kayla ran *k* miles, how many miles did Chen run?

Write an algebraic expression you can use to answer the question.

_____

SRB
200-202

4 Write an equation for each statement. Then solve the equation.

a. 125% of *x* is 625.

Equation: _____

Solution: _____

b. $\frac{1}{4}$ of *y* is 9.

Equation: _____

Solution: _____

SRB
60-64

5 **Writing/Reasoning** Describe how you can use what you know about multiplying and dividing by powers of 10 to help you solve Problem 2.

Generalizing Patterns

Math Message

1 Use the pattern pictured below.

In the table, record numeric expressions for the total number of squares that represent what is shaded and how the pattern is growing.

Plan your numeric expressions to show one part of the expression as constant and the other part as varying.

Hint: The shaded squares may help you see what is constant and what varies.

A sample expression is given for Step 1.

Step Number	Expression for Number of Squares
1	*1 + 3 * 1*
2	
3	
4	

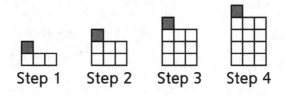

Step 1   Step 2   Step 3   Step 4

2 Describe what is constant and what varies in the pattern.

_____

_____

_____

_____

3 Write an algebraic expression for the number of squares in Step *n*.

_____

Math Boxes

1 For the end-of-year picnic, Jenny will hang a banner from two posts. Gregg measured the distance between the posts as 670.5 cm.

About how many feet of string are needed from post to post?

Number model:

_____

Solution: _____

SRB
32

2 Use the general pattern to compute the following products mentally.

$$x(y + z) = (x * y) + (x * z)$$

Example: $8 * 92 = 8(90 + 2)$
$$= 8(90) + 8(2) = 736$$

a. $5 * 18$ _____

_____

b. $25 * 13$ _____

_____

SRB
204-205

3 Solve using a bar model.

$$3b + 11 = 14 + b$$

Solution: _____

SRB
216

4 Graph the solution set for $-3.5 \geq y$ on the number line.

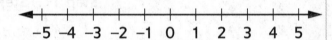

−5 −4 −3 −2 −1 0 1 2 3 4 5

SRB
210-211

5 To keep costs low, the surface area of a tissue box will be less than 122 square inches. Which of the following could be side lengths for the new tissue box if it is shaped like a cube? Circle all that apply.

A.   4.5 inches

B.   4.25 inches

C.   5 inches

D.   4.75 inches

SRB
263-264

6 Use substitution to determine whether the number makes the inequality true or false.

a. $5x > 27$, if $x = 3$ _____

b. $7g + 13 > 26$, if $g = 2$ _____

SRB
210-211

Using Relationships to Make Predictions

Math Message

If you have not already plotted your data on the classroom graphs, follow the instructions on *Math Masters*, page 354 and do so now.

1 Read the Anthropometry section on pages 408 and 409 of your *Student Reference Book*.

2 The following rule is sometimes used to predict the height (*H*) of an adult from the length of the adult's tibia (*t*). Measurements are in inches.

$$H = 2.6t + 25.5$$

Why do you think this rule might not predict the relationship exactly for everyone?

_____

_____

3 Use the rule above to complete the table.

Tibia Length (in.)	Height Predicted (in.)
11	
14	
19	
$17\frac{1}{2}$	

4 Use the prediction line you drew in class to answer the following questions.

    a. The predicted height for a person with a $15\frac{1}{4}$-inch tibia is about _____ inches.

    b. The predicted height for a person with a _____-inch tibia is about 5 feet 0 inches.

5 How closely does the prediction line approximate the actual data points for adult males?

_____

_____

_____

Using Relationships to Make Predictions (continued)

6 How closely does the prediction line approximate the actual data points for adult females? Explain. _____

_____

7 How closely does the prediction line approximate the actual data points for you and your classmates? Explain. _____

_____

_____

8 Scientists can use the same formula and a single bone from a human skeleton to estimate the height of an adult who lived many centuries ago.

   a. The skeleton of a Neanderthal man who lived about 40,000 years ago contained a tibia about $14\frac{3}{4}$ inches long. Estimate the man's height. _____

   b. The tibia of a 20,000-year-old skeleton of an adult was reconstructed and found to be about $12\frac{1}{2}$ inches long. Estimate the person's height. _____

   c. Can you use these heights to draw conclusions about the heights of people 40,000 years ago versus 20,000 years ago? _____

   Explain why or why not.

   _____

   _____

   _____

9 Paul measured his baby sister's tibia (4 inches long) and then used the rule to estimate her height. "That's crazy!" said Paul when he saw the result.

   a. What was Paul's estimate of his baby sister's height? _____

   b. Why might he say the estimate is "crazy"? _____

   _____

   c. Why do you think the formula might result in such a bad prediction?

   _____

   _____

For Problems 10–13, use the graph displaying the wrist and neck data for your class.

10 It is possible to predict a person's neck circumference by doubling his or her wrist circumference. Using N for neck circumference and w for wrist circumference, write an equation representing this relationship. _____

11 **a.** Use the equation you recorded in Problem 10 as a formula to complete the table.

b. How can you draw a prediction line on the class graph?

_____

_____

_____

Wrist Circumference	Neck Circumference
6 in.	
7 in.	
$4\frac{3}{4}$ in.	
$5\frac{1}{2}$ in.	

12 How closely does the prediction line approximate the actual data points?

a. For adult males: _____

b. For adult females: _____

c. For sixth graders: _____

13 Which anthropometry relationship appears to be more accurate, the tibia–height comparison or the wrist–neck comparison?

_____

_____

_____

14 How do you think a formula for predicting anthropometry relationships might be generated?

_____

_____

_____

_____

Math Boxes

① Find the volume of the prism.

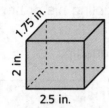

1.75 in.
2 in.
2.5 in.

Number model: _____

Volume: _____

SRB
262

② One dollar bill weighs about $\frac{35}{1,000}$ of an ounce.

To count the money raised at the school dance, the student council weighed it and found it weighed $8\frac{3}{4}$ ounces.

About how much money did they raise if all the money raised was in dollar bills?

Number model: _____

Solution: _____

SRB
68-70

③ Today Rafael is twice as old as his brother Jorge was 3 years ago. Jorge is j years old today.

Write an algebraic expression to represent Rafael's age today.

_____

SRB
200-202

④ Write an equation for each statement. Then solve the equation.

a. 20% of x is 24.

Equation: _____

Solution: _____

b. $\frac{3}{5}$ of y is 3.

Equation: _____

Solution: _____

SRB
60-64

⑤ **Writing/Reasoning** Choose an age for Jorge in Problem 3 and explain how to check whether the expression you wrote makes sense.

Making a Food Budget

Math Message

The spreadsheet below compares the average cost of meals eaten at home versus meals eaten in restaurants for a family of four during one week.

Boxes				⊠
C2 ▼ ○ *fx*	Number of Meals per Week			

	A	B	C	D
1			Food Budget	
2	Home	Cost per Person per Meal ($)	Number of Meals per Week	Total Cost
3	Breakfast	1.00		
4	Lunch	2.00		
5	Dinner	5.00		
6	Restaurant			
7	Breakfast	4.00		
8	Lunch	8.00		
9	Dinner	15.00		
10	Total			

1. How many breakfasts will the family eat in one week (7 days)? _____

2. What is the total number of meals the family will eat in one week? _____

3. Record formulas in the spreadsheet that could do some of the calculations for you as the entries for the Number of Meals per Week column change.

4. What is the least the family can expect to spend for the week? _____

5. The family food budget is $300 per week.

   Use the spreadsheet to make a plan by entering numbers of meals in Column C. Include at least one restaurant meal.

   Fill in the cells in Column C. Then calculate the total cost.
   If you have a spreadsheet program, you can use the program to calculate.
   Otherwise, use your calculator.

Planning a Road Trip

Utah has five national parks: Arches, Bryce Canyon, Capitol Reef, Canyonlands, and Zion.
The Olsen family, a family of four, lives in Salt Lake City and is planning a road trip.
They will start and end at home. They want to visit all of the parks.

In planning the budget for their trip, the Olsens are looking at costs for food, gas, and lodging.
They will be gone for one week.

Food: Use the cost information for home and restaurant meals from the Math Message.

Gas: The Olsens can take their car that gets 30 miles per gallon or their more comfortable van that gets 20 miles per gallon.

Lodging: Each night, they will either camp in one of the national parks or stay in a hotel.

Set up a spreadsheet like the one at the right.

Use formulas whenever possible so calculations change automatically when you change a value in a cell.

In the space below, record at least three formulas you would use. Describe what the formulas calculate.

	A	B	C	D
			Road Trip Budget	
1			Road Trip Budget	
2				
3	Food			
4	Home	Cost/Meal	Number of Meals	Cost
5	Breakfast	$1		
6	Lunch	$2		
7	Dinner	$5		
8				
9	Restaurant	Cost/Meal	Number of Meals	Cost
10	Breakfast	$4		
11	Lunch	$8		
12	Dinner	$15		
13	Food Total			
14				
15	Gas	Cost/Gallon	Number of Gallons	Cost
16	Car	$3.75		
17	Van	$3.75		
18	Gas Total			
19				
20	Lodging	Cost/Night	Number of Nights	Cost
21	Camp	$15		
22	Hotel	$100		
23	Lodging Total			
24				
25	Total			

380

1 Use the map of Utah on *Student Reference Book,* page 353 to decide the order in which they should visit the parks. Record the order below.

1st park: _____   2nd park: _____

3rd park: _____   4th park: _____

5th park: _____

2 Explain how you decided on the order.

_____

_____

3 Use the mileage chart to determine
about how far they will drive.   _____

4 **a.** Explain how you can determine how many gallons of gas the car will use.

_____

_____

b. How many gallons would the car use?   _____

c. How many gallons would the van use?   _____

5 What formula can you use in cell C22 to calculate the
number of hotel nights based on the number of camping nights? _____

6 What information do you enter in your spreadsheet and
where do you enter it if you decide they should use the car?   _____

7 What is the least expensive trip the family can take?   _____

8 **a.** What is the most expensive trip the family can take?   _____

b. Explain how you know it is the most expensive.

_____

_____

9 Use your spreadsheet to design a trip that costs less than $1,000 but includes at least two nights in a hotel and three meals in a restaurant. Copy your solution into the spreadsheet on journal page 380.

Collecting and Analyzing Data

Suppose you are interested in healthy habits of a typical sixth grader. You might ask the following:

- How many glasses of water does a typical sixth grader drink per day?

- How many minutes of exercise does a typical sixth grader get per week?

- How many servings of fruits and vegetables does a typical sixth grader eat per day?

1 Make up your own statistical question about sixth graders' health habits.

_____

_____

2 Collect data on your statistical question and record it below.
Include the units with your measures.

3 Create a dot plot, box plot, or histogram to summarize the results of your data.

4 Describe the distribution of the data you collected.

_____

_____

5 Describe the variation of the data you collected. Use mean absolute deviation or interquartile range to determine the variability in your results.

_____

_____

Math Boxes

1 Geneva recently flew from Chicago, Illinois, to Dublin, Ireland. The total flying time was 7.5 hours. The total distance is about 3,667.5 miles. What was the average speed of the plane?

Number model: _____

Solution: _____

SRB
32

2 Use this pattern to compute products mentally: $x(y - z) = (x * y) - (x * z)$.

Example: $11 * 99 = 11(100 - 1)$
$= 1,100 - 11 = 1,089$

a. $7 * 28$ _____

_____

_____

b. $25 * 99$ _____

_____

_____

SRB
204-205

3 Solve and check.

$$5c + 1 = 7 + 2c$$

Answer: _____

Check: _____

SRB
216

4 Graph the solution set for $d \neq 0$ on the number line below.

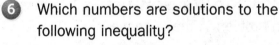
$-4 \quad -3 \quad -2 \quad -1 \quad 0 \quad 1 \quad 2$

SRB
210-211

5 The volume of a cube is 343 cubic inches. What is the surface area of the cube?

Solution: _____

SRB
263-264

6 Which numbers are solutions to the following inequality?

$$15 < 3v - 2$$

Circle ALL that apply.

A. 4

B. 14

C. 6

D. 5.5

SRB
210-211

Math Boxes

383

Math Boxes

1 A rectangular prism has the dimensions 3.5 cm by 4 cm by 6.5 cm. What is the volume of the prism?

Number model: _____

Volume: _____

SRB
262

2 A quarter weighs about 0.2 ounces. Estimate the value of the quarters in the vending machine if they weigh 2 pounds. *Hint:* There are 16 ounces in 1 pound.

Number model: _____

Solution: _____

SRB
68-70

3 Margie's dog weighs 6 pounds more than twice her cat's weight. If her dog weighs *r* pounds, how much does her cat weigh?

Write an algebraic expression you can use to answer the question.

Expression: _____

SRB
200-202

4 Write an equation for each statement. Solve the equation to find the answer.

a. 15% of *a* is 12. What is *a*?

Equation: _____

Solution: _____

b. $\frac{2}{3}$ of *b* is 12. What is *b*?

Equation: _____

Solution: _____

SRB
60-64

5 **Writing/Reasoning** Explain how you can use estimation to check that your answers in Problem 4 make sense.

Notes

Algebra Election Cards 1

Tell whether each is true or false. $10 * x > 100$ $\frac{1}{2} * x * 100 < 10^3$ $x^3 * 1{,}000 > 4 * 10^4$	$T = B - (2 * \frac{H}{1{,}000})$ If $B = 80$ and $H = 100x$, what does T equal?	**Find:** x squared x to the fourth power $\frac{1}{x}$
Evaluate. $x * 10^2$ $x * 10^5$	**Find n.** $n = \frac{(2 * x)}{10}$ $n + 1 = 2 * x$	**Insert parentheses in** $\frac{1}{10} * x - 2$ **so that its value is greater than 0 and less than 4.**	**Find n.** $1{,}000 - n = x$ $1{,}000 \div x = n$
A boulder dropped off a cliff and fell approximately $16 * x^2$ **feet in x seconds.** **How many feet is that?**	**Suppose you earn x dollars per hour. Complete the table.** **Time****Earnings** 1 hr _____ 2 hr _____ 4 hr _____	**Find n.** $n + 10 = 4x$ $n - 10 = 4x$	**Complete.** $x * 10^6 = $ _____ million $x * 10^9 = $ _____ billion $x * 10^{12} = $ _____
Which is less, $\frac{x^3}{10}$ **or** $(x + 10)^2$?	**Which is greater . . .** x^2 or 10^3? x^3 or 10^4?	**What is the value of n?** $n = \frac{(5x - 4)}{2}$	**What is the value of n?** $|-20| + x = n$ $|-100| - x = n$
		What is the value of n? $20 + |x| = n$ $|x| + n = 200$	

Algebra Election Cards 2

Suppose you travel x miles per hour. Complete the table. **Time** **Distance** 1 hr _____ 2 hr _____ 4 hr _____	Compute. $25 + x = ?$ $x + 10 - 3 = ?$	Find a value of n that will make the statement true. $n + 2 > x + 7$ $n - 2 > x - 1$	Is $\frac{1}{x}$ greater than, less than, or equal to $\frac{1}{10}$?
If $(2 * x) + n = 200$, what is the value of n?	What is the median of 4, 8, 12, 12, and x?	What is . . . $x\%$ of 200? $x\%$ of 20?	Name a number less than $\frac{1}{x}$.
You plot triangle EFG with these points: E: (0, 0) F: (x, 0) G: (0, x + 3) How long are $\overline{EG}$ and $\overline{EF}$?	You plot rectangle $ABCD$ with these points: A: (x, 5) B: (x, -5) C: ($-x$, 5) D: ($-x$, -5) What are the lengths of its sides?	Name a number greater than $\frac{1}{x}$.	Tell which is correct for each, $<$, $=$, or $>$? x _____ $300 - x$ x _____ $60 - x$ x _____ $75 - x$
Tell whether it is $<$, $=$, or $>$: $\lvert -5 \rvert$ _____ x $\lvert 30 \rvert$ _____ x $\lvert -2 \rvert$ _____ x	Find w. $w = x^2$ $w = x * 10^0$	If $x + 2n = 200$ ounces, $n =$ _____ ounces	What is n? $5 + 2 * x = n + x$

Spoon Scramble Cards 1

$\frac{1}{7}$ of 42	$\frac{24}{4} * \frac{5}{5}$	$\frac{54}{9}$	$2\frac{16}{4}$
$\frac{1}{5}$ of 35	$\frac{21}{3} * \frac{4}{4}$	$\frac{56}{8}$	$4\frac{36}{12}$
$\frac{1}{8}$ of 64	$\frac{48}{6} * \frac{3}{3}$	$\frac{32}{4}$	$3\frac{25}{5}$
$\frac{1}{4}$ of 36	$\frac{63}{7} * \frac{6}{6}$	$\frac{72}{8}$	$5\frac{32}{8}$

Spoon Scramble Cards 2

$1 \div 2$	$\dfrac{35}{70}$	$\dfrac{1}{8} * 4$	0.5
$\dfrac{1}{3}$	$\dfrac{1}{6} * 2$	$33\dfrac{1}{3}\%$	$\dfrac{1}{2} - \dfrac{1}{6}$
$\dfrac{26}{13}$	$\left(\dfrac{6}{9} * \dfrac{9}{6}\right) * 2$	2	$4 * \dfrac{1}{2}$
$\dfrac{3}{4}$	$\dfrac{600}{800}$	0.75	$3 \div 4$

First to 100 Problem Cards 1

How many inches are in x feet? How many centimeters are in x meters? **1**	How many quarts are in x gallons? **2**	What is the smallest number of x's you can add to get a sum greater than 100? **3**	Is $50 * x$ greater than 1,000? Is $\frac{x}{10}$ less than 1? **4**
$\frac{1}{2}$ of $x =$? $\frac{1}{10}$ of $x =$? **5**	$1 - x =$? $x + 998 =$? **6**	If x people share 1,000 stamps equally, how many stamps will each person get? **7**	What time will it be x minutes from now? What time was it x minutes ago? **8**
It is 102 miles to your destination. You have gone x miles. How many miles are left? **9**	What whole or mixed number equals x divided by 2? **10**	Is x a prime or a composite number? Is x divisible by 2? **11**	The time is 11:05 A.M. The train left x minutes ago. What time did the train leave? **12**
Bill was born in 1939. Freddy was born the same day but x years later. In what year was Freddy born? **13**	Which is larger, $2 * x$ or $x + 50$? **14**	There are x rows of seats. There are 9 seats in each row. How many seats are there in all? **15**	Sargon spent x cents on apples. If she paid with a \$5 bill, how much change should she get? **16**

First to 100 Problem Cards 2

The temperature was 25°F. It dropped x degrees. What is the new temperature? 17	Each story in a building is 10 feet high. If the building has x stories, how tall is it? 18	Which is larger, $2 * x$ or $\frac{100}{x}$? 19	$20 * x = ?$ 20
Name all the whole-number factors of x. 21	Is x an even number or an odd number? Is x divisible by 9? 22	Shalanda was born on a Tuesday. Linda was born x days later. On what day of the week was Linda born? 23	Will had a quarter plus x cents. How much money did he have in all? 24
Find the perimeter and area of this square. x cm x cm 25	What is the median of these weights? 5 pounds 21 pounds x pounds What is the range? 26	$x°$ $?°$ 27	$x^2 = ?$ 50% of $x^2 = ?$ 28
$(3x + 4) - 8 = ?$ 29	x out of 100 students voted for Ruby. Is this more than 25%, less than 25%, or exactly 25% of the students? 30	There are 200 students at Wilson School. x% speak Spanish. How many students speak Spanish? 31	People answered a survey question either *Yes* or *No*. x% answered *Yes*. What percent answered *No*? 32